PRIMITIVE CLASSIFICATION

PRIMITIVE CLASSIFICATION

by

ÉMILE DURKHEIM and MARCEL MAUSS

*Translated from the French
and Edited with an Introduction by*

RODNEY NEEDHAM

THE UNIVERSITY OF CHICAGO PRESS

THE UNIVERSITY OF CHICAGO PRESS, CHICAGO 60637
Routledge and Kegan Paul, Ltd.
© 1963 by Rodney Needham. All rights reserved. Published 1963
Printed in the United States of America

82 81 80 79 78   10 9 8 7 6

International Standard Book Number: 0-226-17332-1 (clothbound)
Library of Congress Catalog Card Number: 63-9737

CONTENTS

INTRODUCTION

I

WHEN A PERSON who has been blind since birth is operated upon and given sight, he does not directly see the phenomenal world which we accept as normal. Instead, he is afflicted by a painful chaos of forms and colours, a gaudy confusion of visual impressions none of which seems to bear any comprehensible relationship to the others. Only very slowly and with intense effort can he teach himself that this confusion does indeed manifest an order, and only by resolute application does he learn to distinguish and classify objects and acquire the meaning of terms such as 'space' and 'shape'.[1]

When an ethnographer begins his study of a strange people he is in a remarkably analogous position, and in the case of an unknown society he may exactly, in no trite sense, be described as culturally blind. He is confronted with a confusion of foreign impressions, none of which can safely be assumed to be what they appear, and the contrast between these and the usages of his own society may be so jarring as even to induce a sense of shock.[2] It is only with

[1] M. von Senden, *Space and Sight: the perception of space and shape in the congenitally blind before and after operation*, translated by Peter Heath, London and Glencoe, Illinois, 1960.

[2] The analogy is precise even to the recognition of a point of 'crisis'. In the case of the patient, this comes when he has made progress in seeing but is suddenly overwhelmed by his relative lack of ability and by the innumerable difficulties which he now realizes he has yet to surmount, when his discouragement may be so severe that he gives up

the most arduous and protracted efforts that he can grasp something of how the people he is trying to understand 'see' themselves and the world in which they live, and not until he has achieved this can he usefully proceed to the technical investigations proper to his academic subject.

The special force of this unique analogy is that it directs our attention to what may be considered the prime and fundamental concern of social anthropology, viz. classification. Evans-Pritchard has well written that 'as every experienced fieldworker knows, the most difficult task in social anthropological fieldwork is to determine the meanings of a few key words, upon an understanding of which the success of the whole investigation depends'.[1] When the ethnographer visits a strange people he carries with him such concepts as 'god', 'power', 'debt', 'family', 'gift', and so on, and however thorough his professional preparation he will tend at first to look for and identify what his own culture denotes by these words and to interpret the statements of the people in terms of them. But gradually he learns to see the world as it is constituted for the people themselves, to assimilate their distinctive categories. Typically, he may have to abandon the distinction between the natural and the supernatural, relocate the line between life and death, accept a common nature in mankind and animals. He cannot pretend to perceive the phenomena involved in any entirely new way, but he can and must con-

---

[1] E. E. Evans-Pritchard, *Social Anthropology*, London, 1951 (p. 80).

and reverts to the dark and tactual world where he has been secure. For the ethnographer, the crisis is that sudden and dismal conviction of ignorance and incapacity by which he is afflicted when he has learned enough to see the complexities of his task but has not yet acquired the felicitous insight which will rescue him from his dejection and revive his resolution.

(Aside from its intrinsic fascination, von Senden's work is full of intriguing parallels with field research which make it worth the attention of any ethnographer.)

ceptualize them in this foreign cast; and what he learns to do in each instance is essentially to classify. Learning the language teaches him to do this in practice, but the language cannot in itself identify the key categories for him or present him with the principles by which they are related. His analytical task, consequently, is first of all to apprehend a mode of classification.

This is the topic of the essay by Émile Durkheim and Marcel Mauss which is presented here. It was first published in 1903, when Durkheim was forty-five and Mauss thirty-one.[1] It is now republished integrally for the first time,[2] and in an English edition, as part of a series of translations of French sociological classics produced under the aegis of Professor Evans-Pritchard by past and present members of the Department of Social Anthropology in the University of Oxford.[3]

It has been selected because of a peculiar combination of theoretical significance and relative academic neglect. The essay is one of the most fascinating and important products of the *Année Sociologique* school, yet it is an odd and perturbing fact that it is virtually unknown to the majority of

[1] 'De quelques formes primitives de classification: contribution à l'étude des représentations collectives', *Année Sociologique*, vol. VI (1901–2), Paris, 1903, pp. 1–72.

[2] Pp. 66–72 of the original have appeared in English translation by Jesse Pitts in *Theories of Society: foundations of modern sociological theory* (edited by Talcott Parsons, *et al.*), Glencoe, Ill., 1961 (vol. II, pp. 1065–8).

[3] Previous publications in the series have been: Émile Durkheim, *Sociology and Philosophy*, translated by D. F. Pocock with an introduction by J. G. Peristiany, 1953; Marcel Mauss, *The Gift*, translated by Ian Cunnison with an introduction by E. E. Evans-Pritchard, 1954; Robert Hertz, *Death and The Right Hand*, translated by Rodney and Claudia Needham with an introduction by E. E. Evans-Pritchard, 1960.

It is planned that the series will continue with the translation by Dr. D. H. P. Maybury-Lewis of Marcel Mauss and Henri Beuchat, 'Essai sur les variations saisonnières des sociétés Eskimos', *Année Sociologique*, vol. IX, 1906, pp. 39–132.

Introduction

professional anthropologists today. Even a symposium such as *African Worlds*, for example, the theme of which is the 'intricate interdependence of social relations and cosmological ideas',[1] nowhere mentions this solitary essay which attempts to demonstrate a constant causal connexion between the two. Sociological commentators on Durkheim's work have also ignored it or have touched upon it only incidentally: Gehlke summarizes the argument, but only to pass on to *Les Formes élémentaires de la Vie religieuse*, of which it is apparently regarded as no more than a forerunner;[2] Alpert similarly mentions it as merely an 'initial formulation' of ideas later worked out in Durkheim's monograph;[3] and Seger excludes its topic from examination, though acknowledging it to be of 'fundamental importance', as being 'too large and at the same time too special'.[4] According to Sommerfelt, it has been overlooked by the majority of linguists as well,[5] and there is convincing evidence of this from quarters where it is least to be expected. The essay deals with the origins and cultural expressions of categories, yet Whorf, to whose interests it is so directly relevant, makes no reference to it;[6] and even the distinguished gathering of linguists, anthropologists, psychologists and philosophers who met in 1953 to discuss Whorf's

[1] *African Worlds: studies in the cosmological ideas and social values of African peoples*, edited with an introduction by Daryll Forde, London, 1954 (p. x).

[2] Charles Elmer Gehlke, *Émile Durkheim's Contribution to Sociological Theory*, New York (Columbia University Studies in History, Economics and Public Law, vol. LXIII, no. 1), 1915 (pp. 46–8).

[3] Henry Alpert, *Émile Durkheim and his Sociology*, New York, 1939 (p. 56).

[4] Imogen Seger, *Durkheim and his Critics on the Sociology of Religion*, New York (Columbia University Monograph Series, Bureau of Applied Social Research), 1957 (p. 4).

[5] Alf Sommerfelt, *La Langue et la Société: caractères sociaux d'une langue de type archaïque*, Oslo, 1938 (p. 9).

[6] Benjamin Lee Whorf, *Language, Thought, and Reality*, edited by John B. Carroll, New York, 1956.

hypotheses about the relationship of linguistic categories to conceptions of the world nowhere mention Durkheim and Mauss's essay in the report on their proceedings.[1]

One reason for this neglect may well be that the essay is buried in a scarce and rather old periodical, but another is distressingly plain, viz. that it is written in French, and the fact that this can be a deterrent to acquaintance with a paper of such quality is the first reason for the publication of an English edition. There are indeed other reasons, as we shall see, by which such neglect might even be thought justifiable, but they would have force only on the part of those already familiar with the essay.

The very fact that the essay resuscitated here is generally ignored, particularly by anthropologists, renders it the more necessary to make a critical examination of its claims to attention today.

II

Durkheim and Mauss concern themselves with symbolic classifications of a moral or religious nature, which they distinguish from practical schemes of distinctions which they call technological. They believe that the human mind lacks the innate capacity to construct complex systems of classification such as every society possesses, and which are cultural products not to be found in nature, and they therefore ask what could have served as the model for such arrangements of ideas. Their answer is that the model is society itself. The first logical categories were social categories, they maintain, the first classes of things were classes of men; not only the external form of classes, but also the relations uniting them to each other, are of social origin; and if the

---

[1] *Language in Culture: conference on the interrelations of language and other aspects of culture*, edited by Harry Hoijer, Chicago, 1954.

Introduction

totality of things is conceived as a single system, this is because society itself is seen in the same way, so that logical hierarchy is only another aspect of social hierarchy, and the unity of knowledge is nothing else than the very unity of the social collectivity extended to the universe.[1] They claim to show how the notion of a system of classification was born, and they conclude by identifying the forces which induced men to divide things as they did between classes.

Their argument is clear, concise, and amply documented; but in logic and method it is open to serious criticism. The most effective and compendious procedure is to set out some of the grounds for criticism here, under these two heads, rather than in frequent and obtrusive footnotes.

The least consequential of the logical flaws are such as are seen at the very beginning of the essay. For example, from the fact that metamorphoses are believed to occur, it by no means follows that definite classificatory concepts are lacking, as Durkheim and Mauss assert.[2] Indeed, the very idea of transmutation could not arise without them; for to believe that a man may change into a parrot one must first have ideas of 'man' and 'parrot' so distinct that a change from one into the other may be conceived at all. And, similarly, to believe in a mystical identity of a man and his totemic animal is not to suffer from 'mental confusion':[3] the individuals involved are nonetheless distinct, only there is a special relationship between the two. Another kind of initial logical error is the inference that the kinship idiom of certain logical classifications shows their social origin;[4] whereas in fact the terms employed ('kinship', 'family', 'genus') themselves merely exhibit a style of classification, and do not in any way indicate (especially since the idiom in this context is so uncommon) an extra-logical origin of the faculty of classification itself.

[1] Below, pp. 82, 83–4.   [2] P. 5.   [3] P. 6.   [4] P. 8.

Introduction

Later in the essay, dealing with ethnographic particulars, Durkheim and Mauss maintain that the position of the 'prey animals' as mediators between the Zuñi and their gods entails that a classification by clans preceded one by quarters;[1] but there is no logical basis for this inference, and it is difficult to imagine how, on the other hand, a spatial region could in itself serve as such a mediator, which is the alternative implied. Or, again, their inference is likewise invalid when they claim that a Zuñi origin myth is 'proof' that in the beginning things were classified by clans and by totems,[2] for the myth may equally logically be claimed to show that originally things were classified by north and south.

Such particular slips are disquieting enough,[3] but they are simply signs of a more general lack of logical character in the argument. At other points we find more explicit examples of a tendency at which the final instances just listed have already hinted. At one, Durkheim and Mauss describe the Omaha system of classification, which divides the universe by 'tribal space' relative to the disposition of clans in the camp and to the route followed on the march.[4] That is, they isolate a form of classification intermediate between a type ordered by clans and one ordered by regions. But instead of being content to do this, or perhaps to emphasize the concordance between social and cosmic divisions which may nevertheless be established, they go on to assert that the systematic idea of regions is only 'in process of formation', and that clans and other things are 'not yet'

---

[1] P. 52.   [2] P. 54.

[3] Durkheim and Mauss contradict themselves also in writing first that the Chinese classification was intended to regulate the conduct of men (p. 71) and that it provided a guide to action (p. 73), and then, to the contrary, that the object of such classifications is 'not to facilitate action, but to advance understanding' (p. 81); but this is a solitary lapse of the kind and not a characteristic logical fallacy such as others dealt with here.   [4] Pp. 55–8.

orientated according to the cardinal points. In other words, even though they lack any evidence of changing modes of classification among the Omaha themselves, they assume that which they intend to establish, viz. that a classification by groups is prior to one by reference to nature. Another case is that in which they consider the division of a tribe into moieties, clans, and sub-clans, in relation to similarly diminishing classificatory categories. There is indeed a formal congruence, but they then assert that this is based on an evolutionary social progression in which moieties are the 'oldest' social groupings and the clans the 'more recent'.[1] Here again, they merely assume a course of development of which they have given no empirical proof, but which it is to the advantage of their thesis to suppose.

This tendency to argument by *petitio principii* is more seriously expressed elsewhere in the essay, beginning with the very first example of classification which Durkheim and Mauss consider. They take a four-section scheme of social classification, by which all the members of a society are comprehensively and integrally categorized, and then abruptly assert that the congruent classification of non-social things 'reproduces' the classification of people.[2] This single word, that is, immediately assumes that which is to be proved by the subsequent argument, viz. the primacy of society in classification. Again, they claim that the astral mythology of certain Australian tribes is 'moulded' by the totemic organization,[3] when all that they have really shown is that stars are so part of a general classification that they may stand in definite relationships to social divisions. In these examples they do not merely assert an evolutionary development in social organization, from the simple to the complex, which makes their argument more plausible, but

[1] P. 83.   [2] P. 11.
[3] P. 29.

they expressly presuppose the very thesis of the argument itself.[1]

Grounds for methodological objection are even more numerous. The first is that in many of the cases which Durkheim and Mauss examine there is a simple lack of correspondence between form of society and form of classification, whereas it is the correspondence which is supposed to make their case. For example, they observe that the Port Mackay marriage classes appear not to have 'affected' cosmological notions.[2] This really means that there is no difference in the forms of classification employed by the societies with moieties previously listed and by this four-section system: in one case the society is divided into two formal groupings, in the other into four, yet both types of society employ an identically dualistic form of symbolic classification. Moreover, the very next society examined, viz. the Wakelbura, is also a four-section system, but in this case the classification does concord with the quadripartite form of social organization.[3]

Durkheim and Mauss consider as a distinct variety of Australian classification that which distributes things not by moieties or by sections but by clans,[4] but the moiety and four-section systems which they examine also have clans, e.g. the Wotjobaluk, who have moieties and at least twelve named clans.[5] Further, the authors regard this latter society as an interestingly complex system in that it distinguishes 'tertiary' divisions in its totemic classification, and they speak of this organization of ideas as being 'parallel'

[1] This feature might be taken to indicate the predominant part played by Durkheim, of whom *petitio principii* may be considered a besetting scholarly vice. Cf. Lévi-Strauss on the same logical fallacy in Durkheim's attempt to establish the collective origin of the sacred (*Le Totémisme Aujourd'hui*, Paris, 1962, p. 102).

[2] P. 12. Note that the very word 'affected' begs the question, as though the four sections must have come first.

[3] Pp. 13–14.   [4] P. 34.   [5] P. 22.

to that of the society; yet the alleged parallelism is not established in this case either, for they cite no evidence that there are actually tertiary descent groups, and the issue is then posed in a yet clearer way by the Moorawaria, who have moieties and a totemic organization embracing no fewer than 152 clans. When Durkheim and Mauss pass on to the Arunta, furthermore, the discrepancy is intensified, for although the form of society is distinct enough (an eight-section system), the Arunta have 'no complete classification, no integrated system'.[1] Now society is alleged to be the model on which classification is based, yet in society after society examined no formal correspondence can be shown to exist. Different forms of classification are found with identical types of social organization, and similar forms with different types of society. Specifically, Durkheim and Mauss's distinction of classification in moiety and section-systems from that in clan-based societies is erroneous, since clans are present in the former as well; so that what we really have are various types of society, some of which classify by moiety, some by section, and some by clan. There is very little sign of the constant correspondence of symbolic classification with social order which the argument leads one to expect, and which indeed the argument is intended to explain.

It is perhaps their most serious methodological failing that Durkheim and Mauss do not subject their thesis to test by concomitant variation. That is, they do not expressly look for societies with identical organization but different forms of classification, or for societies with different organization but similar classifications. Not only this, but when their own evidence presents them with such cases they do not recognize what consequences these must have for their argument. It is not simply that they ignore

---

[1] P. 35.

negative instances, the charge which Evans-Pritchard has laid against the *Année Sociologique* school in general,[1] but that when they do identify such instances they try to explain them away by what he has justly described as 'Durkheim's irritating manœuvre, when a fact contradicts his thesis, of asserting that its character and meaning have altered, that it is a secondary development and atypical, although there is no evidence whatsoever that such changes have actually taken place'.[2] We find the very phrase 'later development' first used in the essay at a point where the presence of a certain clan among the Arunta appears to demand comment,[3] though admittedly there is no particular advantage to the argument in this case. More characteristically, Durkheim and Mauss claim that the multifarious classification by clans among the Arunta is the result of 'changes' in the structure of the society,[4] consequent on the abandonment of the moiety system as the basis of classification, but other than by analogy and on formal grounds no reason is offered to accept that these changes ever took place. It is then simply asserted that if we no longer find a complete classification among the Arunta, this is not because there has never been one but because it disintegrated together with the fragmentation of the clans.[5] Yet there exist other section-systems, with varying numbers of clans, which nevertheless possess integral systems of classification; so that, once more, we are brought to the fact that some societies with sections and clans happen to classify by sections, while other societies of the same type happen to classify by clans. This divergence of practice cannot be nullified by resort to conjectural later developments, but constitutes instead serious evidential objection to Durkheim and Mauss's thesis.

[1] Introduction to R. Hertz, *Death and The Right Hand*, p. 22.
[2] *Ibid.*, p. 12.   [3] P. 31.   [4] P. 35.   [5] P. 37.

The same methodological point arises in the case of the Zuñi, who classify by quarters. Durkheim and Mauss discover disagreements in the various accounts of the distribution of game among the six prey animals, and once more they claim that these can be 'easily explained' by 'modifications' in the orientation of the clans;[1] but they do not offer any such explanation, and they do not specify precisely what modifications took place or what the evidence is that they ever did take place. Their method in this regard is further exposed, and in a particularly revealing manner, in their discussion of the disposition of the Wotjobaluk clans. The figure which they examine[2] is satisfyingly neat and consistent but for the solitary discrepancy that one clan (No. 9) falls to the south of the east–west line by reference to which all the other clans are ordered. They then try to explain this contrary fact away by a variety of expedients: (*a*) there is 'every reason' (though no reason is actually given) to believe that the anomaly is due to an error of observation; (*b*) if it is not such an error, it is due to a 'late alteration' (unspecified) in the original system; (*c*) the informant himself had hesitations on the point; (*d*) there is really no difficulty, since clan 9 is the same as clan 8 (i.e. it is also called Munya), so that it may be considered as not actually falling below the line after all.[3]

The next objection to Durkheim and Mauss's method is that in places they assume that a society employs only one mode of classification at a time. For example, they maintain that the Zuñi system of classification was preceded by one into six regions, and that before this there was one into four, corresponding to the cardinal points.[4] But not only is this conjectured development not demonstrated by the texts to which they refer, it is only plausible on the assumption that the Zuñi could not simultaneously possess

[1] P. 51, n. 3.   [2] P. 61, Figure.   [3] P. 62.   [4] P. 48.

classifications by seven, by six, and by four. Once it is admitted that they might do so, all need to relate these different modes of classification in an evolutionary progression vanishes. In fact, Durkheim and Mauss actually recognize such a possibility in their own account of Chinese classification. This system, they write, is itself composed of 'a number of intermingled systems', a 'multitude of interlaced classifications'.[1] Admittedly, they try to reduce the classification by eight powers into that by the five elements, so that by elimination and combination an exact correspondence of the two shall be effected;[2] but the same is possible in the Zuñi case also. That is, a society need not employ only a single mode of classification; and where it employs two or more the fact that reduction may be possible still does not imply that each distinct classification represents an historical stage in the development of that society's categories.

This issue brings us to the fundamental question of evidence, and to the fact that in a number of places either Durkheim and Mauss do not supply any evidence at all for statements which they make, or the evidence is contrary to their argument. They assert, for example, that the Mount Gambier system of classification shows an increasing differentiation, by clans via moieties, out of a state of initial confusion. Things are divided by moiety, as in the simplest form of classification, but they are also divided by clan within each moiety, and thus resemble more complex systems in which classification is by clans alone without any ordering by moieties.[3] But there is naturally no evidence at all of the primal state of confusion from which the existing, and co-existent, forms of classification are supposed to have emerged, and the whole process is simply assumed by Durkheim and Mauss on purely *a priori* grounds. Concerning the relationship of classification to clans, both types of

---

[1] Pp. 68, 73.   [2] P. 70.   [3] P. 20.

system which Durkheim and Mauss distinguish have (as we saw above) both moieties and clans, so that we are left by this case also with the circumstance that some societies happen to classify by moiety and others by clan, and this remains unexplained. As for the things grouped with one clan being 'undifferentiated' and in a relatively 'amorphous' state,[1] this is merely a gratuitous and implausible elaboration on the part of Durkheim and Mauss themselves, for there is nothing to this effect in the sources they cite, and it is scarcely a credible situation in any case.

The same purely conjectured evolution is presented later in Durkheim and Mauss's flat assertion that 'in a large number of cases' first the moieties were formed and then the clans,[2] whereas neither are the alleged instances specified nor is any one change involved demonstrated factually to have occurred. It is only by means of the mere assumption that complex forms developed from more simple, and the definition of clans as complex and moieties as simple, that they arrive at their conclusion that a classification by clans is a result of change. Similarly, they have no evidence whatever in the reports on the Wotjobaluk for their conclusion that a classification by clans preceded one by spatial regions.[3] Nor is there any evidence that hierarchical classification was based on ideas furnished by the family, clan, and moiety, or that relations between regions were determined by the relative positions of clans.[4] There is a fundamental logical difficulty here to which we shall come below, and one which quite overshadows this objection, but it is important to remark, and even wearisomely to repeat, that no conceivable evidence of any kind is adduced by Durkheim and Mauss to justify such propositions.[5]

[1] P. 20. [2] P. 32. [3] P. 62. [4] P. 66.

[5] Cf. Evans-Pritchard: 'Durkheim and his colleagues and pupils were not content to say that religion, being part of the social life, is strongly influenced by the social structure. They claimed that the religious

Introduction

The role of unevidenced assumption is next illustrated by their statement that it is 'probable' that in China marriage between persons of the same year, or two years of the same name, is regarded as 'particularly inauspicious'.[1] Yet they cite no evidence which would incline one to think that this may be the case, and the sources to which they refer do not justify the supposition. Indeed, by 'probable' they simply mean that it would be a convenient prop to their argument if it were inauspicious, and the addition of 'particularly' merely shows how much they wish it were.[2] It is more serious, though, that they misrender their sources on this point, reading into them things which they do not say. Young does not report that there is a prohibition in Siam on marriage between persons of the same year and animal, and his meaning is perfectly clearly quite different.[3] Nor does Doolittle in any sense even imply that there is a 'quasi-familial' relationship between persons born under the same animal,[4] and that there is consequently a clan-

[1] P. 75.

[2] A similar oddity is their description of the Chinese system of classification as dividing the universe into eight 'families' (*familles*) comparable to the Australian classifications (p. 74), i.e. to classifications by social groups such as clans. It is true that the calendrical cycles are known as the ten mothers and the twelve children (p. 71, n. 3), but the eight classificatory divisions are not known as families. Durkheim and Mauss are trying to establish that the Chinese system is based on the same principles as the Australian and Zuñi classifications, and although they are obliged to recognize that there is no evidence of any connexion between Chinese clans and divisions of space and time, they still seek a social basis for the classification such as they are convinced must have existed. They do not expressly claim that the word 'families' represents a Chinese conception, but in this context it is revealing indeed that they should have chosen this word. We may ponder, too, the influence on themselves of the kinship idiom of European classification which they stress earlier (p. 8).   [3] P. 74, n. 2.   [4] P. 75, n. 3.

conceptions of primitive peoples are nothing more than symbolic representation of the social order. . . . This postulate of sociologistic metaphysic seems to me to be an assertion for which evidence is totally lacking.' (*Nuer Religion*, Oxford, 1956, p. 313.)

like 'exogamy' by classificatory divisions. The usages described in the sources referred to are not 'traces', as Durkheim and Mauss assert, of a classification integrating social with cosmic divisions, and their Chinese evidences on this point lend their argument no support whatever.[1]

Finally, there is Durkheim and Mauss's claim that the 'emotional value' of ideas is the 'dominant characteristic' in classification.[2] This is a profoundly important assertion about a fundamental feature of all human thought, and few propositions could be of more consequence; yet it has to be realized that this factor of emotion is abruptly and gratuitously introduced in this sense only at the end of the paper, and that nowhere in the course of their argument do the authors report the slightest empirical evidence, from any society of any form, which might justify their statement.

Leaving the specific topic of evidence, which has occupied us so far, we have still to examine the part played by sentiment in the argument. The first sign of a resort to affectivity in explaining social facts is seen in Durkheim and Mauss's account of how secondary totems originate, viz. that a group of individuals within a clan come to 'feel' more specially related to certain things which are attributed to the clan in general, so that when the clan becomes too large it tends to split along the lines laid down by the

---

[1] It is a remarkable puzzle that Granet not only does not go into this matter, but even writes that Durkheim and Mauss's few pages on China 'mark a date in the history of sinological studies' (*La Pensée chinoise*, Paris, 1934, p. 29, n. 1). Cf. Robert Merton: 'As Marcel Granet has indicated, this paper contains some pages on Chinese thought which have been held by specialists to mark a new era in the field of sinological studies' ('The sociology of knowledge', *Twentieth Century Sociology*, edited by Gurvitch and Moore, New York, 1945, p. 377). Joseph Needham observes merely that it was 'much more difficult' to explain the Chinese classification by a clear correspondence with society (*Science and Civilisation in China*, vol. II: History of Scientific Thought, Cambridge, 1956, p. 280).

[2] P. 86.

classification.[1] The importance attributed to such sentiment is further underlined in discussing symbolic facts: every divinatory rite, namely, rests on a pre-existing 'sympathy' between certain beings.[2] Later, it is maintained that as logical relations are represented as familial connexions, so they are equally based on the same 'sentiments'.[3] Just what is to be understood here by the assertion that sentiments are the 'basis' of domestic, social, and other kinds of organization is not elaborated, but their determinative importance is repeatedly and increasingly relied upon in the closing pages of the essay. Most notably, we are told that there are sentimental affinities between things as there are between individuals, and that things are classified according to these affinities. This assertion follows from the view that as ideas are systematically arranged for reasons of sentiment (which is now elevated to the status of a finding on which further argument may be premised), so it is necessary that they shall not be pure ideas but shall themselves be products of sentiment. Hence, the differences and resemblances which determine the fashion in which things are grouped are 'more affective than intellectual', and they are differently represented in different societies 'because they affect the sentiments of groups differently'.[4]

It is difficult not to recoil in dismay from this unevidenced and unreasoned resort to sentiment as the ultimate explanation for the complexities of social and symbolic classification whose real significance Durkheim and Mauss have so clearly brought out. But this is the culmination of their argument, and it demands a critical attention proportionate to the importance which they themselves ascribe to it. Their initial premise, from which all else derives, is that social groups are in some way based on sentiment; but, as Lévi-

---

[1] P. 32. Note that this development is entirely conjectured, resting on no reported facts.   [2] P. 77.   [3] P. 85.   [4] P. 86.

Strauss has concisely maintained with respect to ritual,[1] this view rests on a *petitio principii*. Sentiments, intensely though they may accompany aggregation into social groups, are more plausibly the results of such aggregation. They do not explain, in any case, how it is that individuals of common psychic dispositions should engender such systematically different sentiments. More particularly, they do not explain how it is that societies of similar structure, once they are constituted, should attribute to the world such different sentimental values as to compose disparate classifications, and especially when according to Durkheim and Mauss's argument the similarity in social order should incline them towards similar classifications also. Nor, conversely, can a recourse to sentiments illuminate the discrepancies when societies of different structure subscribe to virtually identical classifications. In short, the alleged sentiments explain nothing. It is true, and an important feature of all social life which no sociologist could decline to recognize, that certain ideas may be the objects of intense emotion; but there is neither truth nor use in such an assertion as that space is differentially conceived 'because each region has its own affective value'.[2]

Yet all such particular objections of logic and method fade in significance before two criticisms which apply generally to the entire argument. One is that there is no logical necessity to postulate a causal connexion between society and symbolic classification, and in the absence of factual indications to this effect there are no grounds for attempting to do so. Empirically, Durkheim and Mauss's position is this. Having made their distinction between society and classification, they are confronted by their evidence with a variety of situations: namely, that in societies of similar organization there may be a formal

[1] *Le Totémisme Aujourd'hui*, pp. 102–3.   [2] P. 86.

correspondence of the classification with moieties or with sections or with clans. (The Chinese case may be disregarded, since it exhibits no correspondence at all, and its only value is that it shows that such classifications are not confined to simple societies.) If we allow ourselves to be guided by the facts themselves, i.e. by the correspondences, we have to conclude that there are no empirical grounds for a causal explanation. In no single case is there any compulsion to believe that the society is the cause or even the model of the classification; and it is only the strength of their preoccupation with cause that leads Durkheim and Mauss to cast their argument and present the facts as though this were the case. Indeed, Durkheim had already written that sociological explanation consists exclusively in establishing causal relations.[1] Admitted, Evans-Pritchard has maintained that his attempts to do so are secondary to 'an endeavour to relate the facts to one another in such a way that taken together they are intelligible to us both as a whole and singly',[2] and it may well be concluded that in most of his empirical work this is what Durkheim actually does; but in this essay he and Mauss are explicitly concerned to propound a causal theory, and it is this which they equally evidently fail to establish. Moreover, if such an elucidation were feasible, the indications in the evidence which Durkheim and Mauss used, as well as that from many other parts of the world with which they were certainly familiar, are that the relationship would be the reverse of that which they suppose. That is, forms of classification and modes of symbolic thought display very many more similarities than

---

[1] *Les Règles de la Méthode Sociologique*, chap. 5. It is intriguing to conjecture the effect of nineteenth-century physics on the development of such notions as 'cause' and 'force' in Durkheim's thought, and which led Mauss to look for a 'force' in a gift which compelled its return ('Essai sur le Don', *Année Sociologique*, n.s., vol. I, 1925, p. 33).

[2] Introduction to Hertz, *op. cit.*, p. 15.

do the societies in which they are found; and a causal interpretation, therefore, should rather be that where correspondences between social and symbolic forms are found it
is the social organization which is itself an aspect of the
classification. Actually, in societies practising prescriptive
alliance (such as are typical of Australia) this appears precisely to be the case. But even such systems do not permit
the assertion that the social organization is modelled on, or
reflects, or is caused by the symbolic classification. All that we
are permitted to say is that however we may divide the social
ideas in question (into 'social order' and 'symbolic order',
for example, or into more numerous analytical categories
of facts), they exhibit certain common principles of order,
no one sphere of interest being the cause or model of the
organization of another. Whatever the empirical validity or
analytical value of this interpretation, it is a logical alternative to their causal analysis which Durkheim and Mauss
never consider.

The second point of general criticism is the more serious,
since it shows Durkheim and Mauss's entire venture to
have been misconceived. They aptly call their essay a 'contribution to the study of collective representations', but
their real concern throughout is to study a faculty of the
human mind. They make no explicit distinction between
the two topics, and indeed they argue as though there
were none to be made, so that conclusions derived from a
study of collective representations are taken to apply
directly to cognitive operations.

The failure to make this essential distinction was noted
nearly fifty years ago by Gehlke with regard to *Les Formes
élémentaires de la Vie religieuse*, when he observed that
Durkheim saw the categories as 'a content of mind rather
than as a capacity of mind', and that this was 'quite consistent with Durkheim's conception of the mind as a system

xxvi

of representations, rather than as a functioning whole'.[1] Some years later, Dennes elaborated this expository comment into a cogent criticism of Durkheim's work on religion which applies with equally invalidating effect to Durkheim and Mauss's main argument. As he writes, 'Durkheim's theory of the origin of the categories depends on his ambiguous conception of mind'.[2] If the mind is taken to be a system of cognitive faculties, it is absurd to say that the categories originate in social organization: the notion of space has first to exist before social groups can be perceived to exhibit in their disposition any spatial relations which may then be applied to the universe; the categories of quantity have to exist in order that an individual mind shall ever recognize the one, the many, and the totality of the divisions of his society; the notion of class necessarily precedes the apprehension that social groups, in concordance with which natural phenomena are classed, are themselves classified. In other words, the social 'model' must itself be perceived to possess the characteristics which make it useful in classifying other things, but this cannot be done without the very categories which Durkheim and Mauss derive from the model.

If, on the other hand, the mind is taken to be simply a collection of ideas, varying from culture to culture, the study of these ideas can never expose the origin of those fundamental categories of the human mind by which, in every culture and at every period, they are themselves universally ordered. Different peoples conceive space and time differently, but no comparative study of their concepts can yield the origin of the categories of space and time; they classify by different principles, but in no circumstances can

[1] Gehlke, *op. cit.*, p. 53.
[2] William Ray Dennes, *The Method and Presuppositions of Group Psychology*, Berkeley (University of California Publications in Philosophy, vol. 6, no. 1), 1924 (p. 39).

the study of these show how the faculty of classification itself originated.

Durkheim and Mauss are led by this ambiguous conception of mind to assert that the individual mind is incapable of classification, and their venture as they conceive it derives much of its justification from this assumption. Now no one would pretend that the individual could ever construct, without education in the categories of his society, a complex classification of collective representations such as the society has inherited from a long history. But this in no way implies that the individual mind lacks the innate faculty of classification; and it would be difficult to conceive, in any case, how an individual might even apprehend a classification unless the mind were inherently capable of the essential operations by which classes are constituted. Even on this score, moreover, Durkheim and Mauss lose their case by conceding too much. They admit, that is, that a developing consciousness distinguishes right from left, past from present, that it perceives resemblances, can separate the one from the many, and can group things.[1] Yet consider how formidable an apparatus of concepts is already presumed: space, time, quantity, and in fact the very ability to classify. They say that this is about all that even an adult mind could produce without education; but when so much is admitted, what in principle is there left to challenge? Yet even such a criticism is not fundamental enough. The developing consciousness in these respects is never observed in an individual unaffected by education, and the 'rudimentary distinctions' observable in a child are themselves collective representations inculcated by instruction. They tell us

[1] P. 7. They are also prepared to admit that mankind has always employed practical classifications of means of subsistence (p. 81, n. 1). Cf. E. Benoit-Smullyan, 'The sociologism of Émile Durkheim' (*An Introduction to the History of Sociology*, edited by H. E. Barnes, Chicago, 1948, pp. 499–537), p. 532, n. 61.

nothing about an innate incapacity to classify, but instead demonstrate an innate capacity to learn to classify.

We have thus to conclude that Durkheim and Mauss's argument is logically fallacious, and that it is methodologically unsound. There are grave reasons, indeed, to deny it any validity whatever.

III

It might be asked whether there is really any point in republishing, with all the care demanded by translation and editing, a work which is so seriously defective in so many respects; but this would be to misconceive the purpose of an English edition and the nature of sociological understanding. A critical introduction cannot be hagiography, and the intellectual value of an argument does not depend solely on its validity.

Durkheim and Mauss's essay is in fact still singularly worth reading for its historical, methodological and theoretical interest. Its historical interest is, to begin with, that it is an early formulation of ideas later more famously expressed in Durkheim's *Les Formes élémentaires de la Vie religieuse* (1912). Parts of the essay were recapitulated in the latter work, of which there has long been an English translation,[1] but only a very few pages reproduce material from the essay, and then only in the form of summary examples which give little idea of the scope or method of its argument. So an English edition of 'De Quelques formes primitives de classification' not only serves to show a wider public the early development of certain of Durkheim's most renowned ideas,[2] but conversely displays the empirical and

[1] *The Elementary Forms of the Religious Life*, translated by Joseph Ward Swain, London [1915].

[2] There is no indication in the essay itself of how much of the argument is due to Mauss, and Durkheim in his monograph gives none. It may be noted, however, that in the year before the essay appeared,

Introduction

analytical grounds on which their subsequent expression was based.

The first world war nearly destroyed the *Année Socio-logique* school, and tragically cut short the lives of young scholars who were taking up particular problems in classi-fication, such as Antoine Bianconi, who had embarked on a study of categories in Bantu languages before he was killed in 1915;[1] but the essay itself continued in other ways to exercise an influence which amply secures its title to a place in the history of sociological thought. We may single out two lines of development as the most prominent.

The essay had its most marked and continued effect through Robert Hertz, a pupil of Durkheim's, also killed in action in 1915, who was clearly inspired by it to write his article on the pre-eminence of the right hand.[2] In this, Hertz examines dualistic forms of symbolic classification which are associated with right and left, and he attempts to explain their common characteristics by reference to a prin-ciple of dualism fundamental both to thought and to primitive forms of social organization. The influence of

[1] Marcel Mauss, 'In Memoriam: l'œuvre inédite de Durkheim et de ses collaborateurs', *Année Sociologique*, n.s., vol. I, 1925 (pp. 22–3).

[2] 'La prééminence de la main droite: étude sur la polarité religieuse', *Revue Philosophique*, vol. LXVIII, 1909, pp. 553–80. (English trans-lation in *Death and The Right Hand*, London, 1960.)

Henri Hubert had already published, in a review section which he produced in collaboration with Mauss, certain seminal observations on time and space which are direct intellectual precursors of the essay on classification (*Année Sociologique*, vol. V, 1902, p. 248) and which he and Mauss later developed into an essay on the religious, i.e. social, origin of the concept of time (H. Hubert, 'Étude sommaire de la re-présentation du temps dans la religion et la magie', École Pratique des Hautes Études, Section des Sciences Religieuses, Paris, 1905, pp. 1–39; cf. H. Hubert and M. Mauss, *Mélanges d'Histoire des Religions*, Paris, 1909). But perhaps this kind of search into the origins of these ideas is misdirected, for as this very example indicates the *Année Sociologique* school themselves composed a scholarly *conscience collective* characterized by a remarkable co-operation and unity of thought.

Introduction

Durkheim and Mauss's essay, which he cites, is evident throughout, and especially in his conclusion that 'the intellectual and moral representations of right and left are true categories, . . . since they are linked to the very structure of social thought'.[1] Hertz's paper inspired in its turn a series of investigations into forms of symbolic classification connoted by right and left, in China, Celebes, Greece, and elsewhere, of which two of the more recent, concerning Africa, are E. E. Evans-Pritchard's 'Nuer spear symbolism',[2] and T. O. Beidelman's 'Right and left hand among the Kaguru: a note on symbolic classification'.[3] Interest in this form of classification continues,[4] and it is to Durkheim and Mauss's essay that the realization of its significance may ultimately be traced.

The essay made a most notable impact on social anthropology in the Netherlands. There, the Leiden school produced an impressive body of studies, by scholars such as F. D. E. van Ossenbruggen, J. P. B. de Josselin de Jong, and W. H. Rassers, in which the connexion with Durkheim and Mauss's work was expressly recognized; and the influence may easily be discerned in other and later publications where the intellectual genealogy is not made explicit.[5] Some of these works were of special importance, too, in that ideas elaborated on the basis of material from Australian aborigines were applied in them to the study of Indonesian societies of high civilization.

[1] *Death and The Right Hand*, pp. 112–13.

[2] *Anthropological Quarterly*, n.s., vol. I, 1953, pp. 1–19. (Reproduced as chap. 9 of *Nuer Religion*, Oxford, 1956.)

[3] *Africa*, vol. XXXI, 1961, pp. 250–7.

[4] A handbook bringing together a collection of essays on dualistic symbolic classification with special reference to right and left, including the publications referred to, is in preparation at Oxford.

[5] A detailed survey of the work of the Leiden school is a task of considerable importance in the history of social anthropology, though not to be undertaken here.

Introduction

An impression of the historical significance of the essay in the development of the discipline may be gained by listing some of the studies to which it has proved relevant or on which it has exercised an acknowledged theoretical influence. These include: 'De oorsprong van het Javaansche begrip Montjå-pat', by F. D. E. van Ossenbruggen;[1] 'Quelques particularités de la langue et de la pensée chinoises', by Marcel Granet;[2] 'Over den zin van het Javaansche drama', by W. H. Rassers;[3] 'De oorsprong van den goddelijken bedrieger', by J. P. B. de Josselin de Jong;[4] *La Pensée chinoise* by Granet;[5] *La Langue et la Société: caractères sociaux d'une langue de type archaïque*, by Alf Sommerfelt;[6] 'La Pensée cosmologique des anciens Mexicains: (représentation du monde et de l'espace)', by Jacques Soustelle;[7] *Ormazd et Ahriman: l'aventure dualiste dans l'antiquité*, by J. Duchesne-Guillemin;[8] 'Social structure', by Claude Lévi-Strauss;[9] *The Structure of the Toba-Batak Belief in the High God*, by Ph. L. Tobing;[10] *Science and Civilization in China* (vol. II: History of Scientific Thought), by Joseph Needham;[11] *Anthropologie Structurale*, by Lévi-Strauss;[12] and a number of more recent analyses in social anthropology.

The stature of these authors and the renown of their works are enough to indicate the value of Durkheim and Mauss's essay; and the regularity with which it has been

[1] *Verslagen en Mededeelingen der Koninklijke Akademie van Wetenschappen*, Afdeeling Letterkunde, 5e reeks, 3e deel, 1918, pp. 6–45.

[2] *Revue Philosophique*, vol. LXXXIX, 1920 (p. 188).

[3] *Bijdragen tot de Taal-, Land- en Volkenkunde van Nederlandsch-Indië*, vol. 81, 1925 (p. 359).

[4] *Mededeelingen der Koninklijke Akademie van Wetenschappen*, Afd. Letterkunde, deel 68, serie B, 1929 (p. 6).

[5] Paris, 1934 (p. 29).   [6] Oslo, 1938 (pp. 9–13).

[7] *Actualités scientifiques et industrielles*, 881 (Ethnologie), Paris (p. 6).

[8] Paris, 1953 (p. 86).

[9] *Anthropology Today*, edited by A. L. Kroeber, Chicago, 1953 (p. 532).   [10] Amsterdam, 1956.

[11] Cambridge, 1956 (pp. 279–80).   [12] Paris, 1958 (pp. 8, 362).

called upon, in very different works, over more than fifty years shows its fundamental relevance and its continued power of inspiration.

The methodological interest of the essay might be **thought** predominantly negative, but there is a great deal to be learned from Durkheim and Mauss's very mistakes. After all, they were two masters of the subject, working in collaboration on a practically unexplored topic of great importance, and an intellectual exercise of this kind could not fail to be instructive. It should not be forgotten, either, that what may readily be perceived as mistaken lines of enquiry today could not have been nearly so evident at the turn of the century. Scholars of the day were powerfully constrained by the prevailing style of thought to analyse human affairs causally and historically, and it would be unreasonable to expect Durkheim and Mauss, prescient though they may appear in some respects, and however recalcitrant to such analysis their evidence may appear to us, to have departed widely from such aims. In any case, these remain respectable aims in social anthropology, and it is merely our advantage that they have been shown not to be the only ones conceivable, and that we can more easily recognize when they are likely to be unrealizable.

The more obviously positive methodological interest of the essay is also considerable. It shows the great and indispensable advantages, as Evans-Pritchard has elsewhere pointed out in another connexion, of 'an intensive study of a limited and clearly defined cultural region where the facts can be examined in their full contexts of ideas and practices'.[1] Social anthropology could not confine itself to this procedure and still pursue its just scholarly ambitions, but repeated contrasts with Frazerian comparative method and

[1] Introduction to Hertz, 1960, pp. 14–15.

with more recent statistical comparisons have effectively demonstrated that this is how we have to begin, at all events, in any attempt at sociological explanation. Another methodological procedure, later to become characteristic of the *Année Sociologique* school, of which the essay gives an example is that of seeing a certain range of facts in their totality, as composing in a systematic fashion a whole of which the parts cannot be adequately comprehended in isolation from each other. As a precept, this may seem obvious and even trite today, but it is not at all obvious that a systematic relationship may be established between a rule of marriage and the attribution of responsibility for a death,[1] between the route followed by a people on the move and the cosmological associations of its component groups.[2] In any case, however readily conceded in principle, it would by no means be difficult to cite investigations today which in practice suffer from lack of attention to the precept. A last virtue of Durkheim and Mauss's essay which deserves particularly to be mentioned is the sheer ingenuity of their argument. Given their premises, it is by most standards a remarkable piece of work in construction and in clarity of exposition, and these are qualities not to be taken for granted in any period.

The theoretical significance of the essay, finally, secures it a permanent place as a sociological classic. Its great merit, and one which outweighs all its faults, is that it draws attention, for the first time in sociological enquiry, to a topic of fundamental importance in understanding human thought and social life.

The importance of this notion of classification is most easily illustrated from the study of descent systems. Kroeber stated the case in 1917, when he wrote: 'All words necessarily classify according to certain principles which usually

[1] Pp. 14–16. [2] Pp. 57–8.

Introduction

are not more than half conscious. There is no conceivable reason why terms of relationship should be an exception, and no evidence that they are.' [1] Or, as he later phrased the issue: 'Every kinship system is . . . a little system of classificatory thought.' [2] Hocart made the same point in his masterly essay on kinship systems,[3] where he urged and demonstrated that terminologies of relationship should be approached primarily through the categories and principles of classification employed by the people themselves. An outstandingly effective analysis in this regard is E. R. Leach's 'Concerning Trobriand clans and the kinship category "tabu" ',[4] in which he takes the kinship terms, so far as possible without theoretical presuppositions, as 'category words' which have to be understood in relation to the social structure, and the culmination of which is his convincing suggestion that in his analysis he has come close to the Trobrianders' own conception of their meanings. In case after case, as other recent analyses may also demonstrate, it becomes possible to elucidate otherwise anomalous and puzzling descent systems simply by looking on them as forms of social classification, and then, by an imaginative apprehension of their categories, discerning their principles of articulation. This done, and such a basic understanding having been acquired, it then becomes more feasible and more profitable to proceed with technical or theoretical

[1] A. L. Kroeber, 'California kinship systems', *University of California Publications in American Archaeology and Ethnology*, vol. 12, 1917 (p. 390).

[2] 'Kinship and history', *American Anthropologist*, vol. 38, 1936 (p. 339). It may be remarked in this connexion, since the article is part of an exchange with Radcliffe-Brown, that Kroeber showed himself, here and in his empirical analyses, a considerably more perceptive interpreter of descent systems than the latter, who never really grasped the necessity to view a kinship terminology as a form of classification.

[3] A. M. Hocart, 'Kinship systems', *Anthropos*, vol. XXXII, 1937, pp. 245–51. (Reprinted in *The Life-Giving Myth*, London, 1952.)

[4] *Cambridge Papers in Social Anthropology*, no. 1, 1958, pp. 120–45.

exercises in analytical terms foreign to the classifications themselves.

In a wider context, there are the splendid works of Georges Dumézil on the common forms of early Indo-European social and religious classification,[1] and his fascinating examination of Indo-European ideas of dual sovereignty as expressions of a 'principle of classification'.[2] F. D. K. Bosch's pioneering analysis of Indian symbolic motifs in relation to ancient cosmic classifications[3] shows yet more clearly that the relevance of Durkheim and Mauss's theme is not confined to the field of primitive social organization.

It is not claimed that all these scholars were directly inspired by the essay on primitive classification, though in another place it might well be shown that some at least were (Dumézil, for instance, dedicates *Mitra-Varuna* to Mauss and Granet), and this is irrelevant in any case to the issue of indicating the importance of the concept of classification itself.

Its importance is further shown, in a simple and direct fashion, by the fact that it is still possible usefully to investigate the relationship, postulated by Durkheim and Mauss, between symbolic classification and social structure. This varies, not only with the general complexity of culture of which it is a part, and with the state of scientific and technological advance, but also with the type of descent system.

[1] A short account of some of his investigations may be found in *Les Dieux des Indo-Européens*, Paris, 1952. See also *L'Idéologie tripartie des Indo-Européens*, Brussels (Collection Latomus, vol. XXXI), 1958.

[2] *Mitra-Varuna: essai sur deux représentations indo-européennes de la souveraineté* (2e édition), Paris, 1948 (p. 206). For a social anthropological analysis of classification and complementary sovereignty, inspired by Dumézil's work as well as by Durkheim and Mauss's essay, see: 'The Left Hand of the Mugwe: an analytical note on the structure of Meru symbolism', *Africa*, vol. XXX, 1960, pp. 20–33.

[3] *De Gouden Kiem*, Amsterdam-Brussels, 1948; revised English edition, *The Golden Germ: an introduction to Indian symbolism*, 's-Gravenhage, 1960.

Recent investigations make it appear, namely, that in cognatic societies the relation of symbolic to social order may be insignificant or minimal, that in simple lineal descent systems the relationship may be discerned in a range of particulars or in isolated institutions but not usually in any comprehensive manner, and that in systems of prescriptive alliance there is such a concordance between the symbolic forms and social organization that these two orders of facts may be regarded as aspects of one conceptual order, one mode of classification. This concordance need not be a formal correspondence, such as Durkheim and Mauss supposed, but may subsist in a structural sense, institutions of different forms being seen as based on the same mode of relation. Thus societies with moieties, section-systems, and systems of asymmetric alliance vary considerably in form and may appear disparately ordered, yet when analysed in terms of their component dyadic relation they may all be seen to agree with the dualistic schemes of symbolic classification which they characteristically possess. On the basis of such studies, moreover, it may be understood how it was that, simply by selecting Australian societies as their supposedly typical cases, Durkheim and Mauss were predisposed to assert a general correspondence of the kind in other simple societies and to look for it where it was not to be found, for the majority of Australian societies practise prescriptive alliance and therefore, it may be argued, present striking concordances between social organization and symbolic classification which are actually uncommon in simple societies in general.[1]

The importance of classification may further be indicated by glancing very briefly at a number of problems which

[1] It may be, too, that this is a major reason why Radcliffe-Brown, whose theoretical concern was primarily with Australian societies, emphasized so strongly the unity and institutional harmony of primitive societies in general.

may be seen in its light as belonging to a common class. For instance, it is an extremely common feature of human societies, and especially of those in which jural status is defined primarily by descent, that they govern marriage and sexual intercourse by the strictest of rules. Infringement of these rules, usually known as incest, is regarded as especially heinous and may even be punishable by death. But it is a surprising fact that in a number of societies, and even in those whose descent systems might seem to make it inconceivable, the people or certain groups or offices are said in myths to be descended from a primal act of precisely such incestuous intercourse as is actually most abhorred. In India it is a black sin to sleep with one's daughter, yet in one prominent myth of origin the creator, Prajāpati, engenders the human race on the body of his own daughter; certain Eskimo believe they are descended from the union of brother and sister; a Sumbanese clan traces its origin to an act of bestiality with a dog, something so vile that even in the myth the perpetrator is shamed into suicide.[1] Examples of the kind could be multiplied with ease, and the most cursory survey of ethnography from many areas of the world yields similar instances. Such myths vary greatly in their particulars, but they all exhibit a common feature, viz. that the principles of the extant social classification are inoperative or are flouted. The myth, that is, may represent the present order as emerging from a primeval chaos, or it may reverse the relations between categories having a primeval definition; but in either case the problem has to do with classification.

Circumstances of the kinds delineated in such myths are

[1] No references are given for examples adduced here, since they are merely reminders, having no special value in this context, of very general and well-known features of human society. Their interest, that is, lies not in their cultural singularity but in the classificatory processes, far less well recognized, which they illustrate.

not only imagined, but are realized in social life, in periodic enactments usually referred to as saturnalia. Here again there are two kinds of manipulation of the categories of the social order, parallel to what may be discerned in myth. In one, there is a period of chaos in which all relationships are confounded: rules of incest, property, and social etiquette are temporarily abrogated. In the other, relations between social categories are strictly reversed, so that a master obeys a slave in his own household, an officer waits on his men at table. But both kinds of institutionalized disorder, so wide-spread throughout the world and at all periods of history, are for the social anthropologist problems in classification.

The theme of reversal is itself one of the most pervasive and fundamental problems in social anthropology, and it is so only in the context of the classifications within which its instances are discerned. The most general incidence of symbolic reversal is to be found in rites of which certain kinds of saturnalia may be examples, in the attribution to the usually despised and inauspicious left hand of certain special virtues usually associated with the right, in the employment of homosexual or impotent men as priestesses. Transvestism, of which this last institution is itself an example, is a particularly intriguing kind of reversal, as when Bornean women at an agricultural rite dress themselves as warriors in all the paraphernalia, usually forbidden to them, of headhunting. Another form of symbolic reversal, and an especially important one, is that used to mark a boundary, between peoples, between categories of persons, between life and death. Hostile or suspect neighbours of the Lugbara are inverted; witches among the Kaguru dance upside down; in the Toraja land of the dead everything is the reverse of what it is in this world, to the extent that words even mean the opposite of their everyday connotations or are pronounced backwards.

Introduction

All these examples involve relations between categories, i.e. they are problems in classification. They are of outstanding importance, for if our first task as social anthropologists is to discern order and make it intelligible, our no less urgent duty is to make sense of those practically universal usages and beliefs by which people create disorder, i.e. turn their classifications upside down or disintegrate them entirely.

The focus of social anthropology is order, and this is secured and denoted by systematically related categories, i.e. by classification. As Durkheim and Mauss write in a more specific context, classifications 'express . . . the very societies within which they were elaborated'.[1] It is the paramount importance of this topic which lends their essay the singular value that it still possesses. Whatever its faults, its prime theoretical contribution has been to isolate classification as an aspect of culture to which sociological enquiry should be directed.[2]

While it may readily be conceded that empirically classification is important enough, it may be doubted whether simply the notion of 'classification' is likely to be very consequential in analysis. But this would be to ignore what is perhaps the most significant lesson in the development of sociological thought in the last sixty or more years, i.e. since the establishment of sociology in France as an empirical discipline. Evans-Pritchard has written, referring to members of the *Année Sociologique* school: 'It is a fact, which none can deny, that the theoretical capital on which anthropologists today live is mainly the writings of people whose research was almost entirely literary.' [3] Now this theoretical capital does not consist, whatever the announced aims of its

[1] P. 66.

[2] Cf. Professor Evans-Pritchard on the necessity to return to the study of 'primitive philosophies' (*Nuer Religion*, pp. 314–15).

[3] Introduction to Hertz, 1960, p. 24.

creators, of a collection of sociological laws, general theories, and more specific abstract propositions, all linked together into a coherent body of theory. Sociological laws of functional interdependence have not yet been established in social anthropology,[1] no general theory has so far emerged, and a succession of testable hypotheses have led (where they have led anywhere) not to abstract formulae of social life but to empirical generalizations. Rather than now possessing a solid theoretical basis of this kind, social anthropology is in a state of conceptual confusion expressed in proliferating technical taxonomies and definitional exercises, each new field study offering enough 'anomalous' features to lead to yet more typological and methodological pronouncements.[2] We have reached a point of empirical plenitude and propositional futility at which Leach's precept that in anthropological analysis 'we must take each case as it comes'[3] inspires both relief and hope. As he persuasively writes, 'ethnographic facts will be much easier to understand if we approach them free of *all . . . a priori* assumptions. Our concern is with what the significant social categories are; not with what they ought to be.'[4] With these words we are back to Durkheim and Mauss, back to 1903.

Yet this appears strange, for social anthropology in the same period has actually made most encouraging and relatively rapid progress in describing, analysing, and rendering more intelligible a range of types of society and collective representations whose variety and complexity increase with each expedition and each augmentation of professional skill.

[1] Evans-Pritchard, *op. cit.*, p. 14.

[2] This conclusion is not a matter for any satisfaction, a partial view inspired by an anti-scientific or, worse, 'literary' attitude to social anthropology. It is a matter of fact, evident to anyone who has reflected on it—or, if it is not, then no one has so far established the contrary.

[3] E. R. Leach, *Rethinking Anthropology*, London (London School of Economics, Monographs on Social Anthropology, no. 22), 1961 (p. 10).

[4] *Ibid.*, p. 27.

Introduction

We have only to compare *Nuer Religion* with *The Andaman Islanders*, *Political Systems of Highland Burma* with *The History of Melanesian Society*, *Une Sous-caste de l'Inde du Sud* with *The Veddas*, or *Les Structures élémentaires de la Parenté* with *Systems of Consanguinity and Affinity*, to be convinced of a real progress, and to be inspired, perhaps, with a restored faith in the value of the subject. How, then, has this come about, and what is the theoretical basis for these advances?

Partly this progress is the result of increasingly rigorous standards of fieldwork, and a vast accumulation of reliably ascertained facts. We certainly possess a far more extensive and minute factual knowledge than did Durkheim and Mauss about what human beings in society do, a knowledge sometimes expressible, more or less precisely, in empirical generalizations. These generalizations in turn sometimes permit the formulation of specific propositions which are susceptible to empirical test, and they sometimes lead to ethnographic or analytical advances of a technical kind. But there is more to the matter than this. Social anthropology in Britain (to speak only of that country where it has acquired most renown in recent decades) has been inspired by certain general ideas, subtly derived from the early French sociologists, which have had a singular theoretical influence, and much of the progress is to be attributed to them.

They are analytical notions such as 'transition', 'polarity' (opposition), 'exchange', 'solidarity', 'total', 'structure', 'classification'. Now these are not theories but highly general concepts; they are vague, they state nothing. At first sight there is nothing to be done with them, and certainly they cannot be taught as elementary postulates in introductory courses of social anthropology. Indeed, their significance is only apprehended after arduous application to the task of understanding social phenomena; the less one knows about human society and collective representations the less they appear to mean. Yet they have proved to possess a great and perennial analytical value, such that it may be claimed that it is they which are essentially the 'theoretical capital' of social anthropology.

Their generality and their practically indefinable significance, however, do not at all preclude precise formulation of problems or useful conclusions. On the contrary, once Mauss, for instance, had established the sociological significance of exchange,[1] Lévi-Strauss was able to construct a rigorous scheme of analysis covering types and modes of exchange, and their differential social correlates, in a sphere where the notion might have seemed least applicable, viz. the regulation of marriage and the problem of 'incest'.[2] His classical monograph, in turn, has led to a number of increasingly technical studies in prescriptive alliance, but whatever their methodological refinements, and however precise the resulting generalizations, their theoretical cogency stems primarily from the simple notion of 'exchange'.

Such analytical notions, whatever they may lead to, seem in themselves not to be capable of elaboration or application such as leads to the formulation of general theories in the exact sciences. Certainly, in spite of their fertility, the theoretical progress which they have facilitated has not in fact consisted in the construction of general theories or laws. But they are by no means to be under-rated simply because they are not abstract propositions of this kind: on the contrary, they have achieved gratifying success in rendering many aspects of social life intelligible. Nevertheless, it has to be contemplated as a possibility that this kind of enquiry is all that we may ever succeed at in social anthropology,

[1] 'Essai sur le Don', 1925.
[2] Claude Lévi-Strauss, *Les Structures élémentaires de la Parenté*, Paris, 1949.

and that such analytical notions may prove to be perennial in more than a figurative sense. It is possible, that is, that they are unique and true insights into social life and collective representations anywhere, in other words that they are categories of sociological thought.

It is the cardinal achievement of Durkheim and Mauss's essay, with all its imperfections, to have conceived the analytical notion of 'classification' in sociological enquiry. However we may turn the notion to our scholarly purposes, it has set us the urgent task so well set out by Mauss: 'We must first of all draw up as large as possible a catalogue of categories, beginning with all those which it can be discovered that mankind has ever employed. It will then be seen that there have been, and that there still are, many dead moons, and others pale or obscure, in the firmament of reason.' [1] If social anthropology had no other aim than this, it would be the grandest of enterprises in human understanding.

IV

Now that some of the grounds for producing an English edition of this essay have been summarily examined, there may be occasion for a comment on the place of translation in general in a social anthropologist's scheme of duties.

Very few academics have any ideas of their own, and their livelihood largely consists in handing on the teachings of the masters of their subject. Some of these teachings are compendiously set out in specific disquisitions on various topics, and if these are in foreign languages then it is a

[1] More striking and evocative in the French of which this is an inevitably flat rendering: 'Il faut avant tout dresser le catalogue le plus grand possible de catégories; il faut partir de toutes celles dont on peut savoir que les hommes se sont servis. On verra alors qu'il y a eu et qu'il y a encore bien des lunes mortes, ou pâles, ou obscures, au firmament de la raison' (*Sociologie et Anthropologie*, Paris, 1950, p. 309).

highly useful academic activity not only to lecture on them but to translate them as well. University teachers may be assured that it is no great burden to do so, and that there is moreover a positive and often pleasurable advantage to be found in it. It is very profitable to pore with intense concentration over the writings of a great scholar, both for the increased sympathy with his thought (and the humility) which it brings,[1] and for the novel considerations in one's own work to which this may lead. To this may be added, also, the satisfaction engendered by a consciousness of being in an intellectual tradition, of being related in an ideological sense to one's great predecessors, and this is considerable.

Something of these benefits can be presented to as many students as will come newly to the work through the translation, and this is surely an academic service and a source of gratification. It is true that there is little professional credit to be gained, for any social anthropologist should be competent to make a translation from the French, for example, and most will think (some correctly) that they could have done it better, but this can scarcely be a dominant concern to a scholar, and especially when other academic gains are so considerable.

The present translation preserves the form of the original text, in that Durkheim and Mauss's punctuation and their divisions into paragraphs and sentences have been closely adhered to. For the most part, these are not due to distinctive characteristics of French prose, or to contemporary literary fashion, but have expository value: that is, they reflect something of the way in which Durkheim and Mauss brought together in their minds the material they used, and the fashion in which they thought their argument

[1] Cf. Durkheim: 'If you wish to mature your thought, apply yourself scrupulously to the study of a great master; dismantle a system down to its most secret workings'. (Harry Alpert, *Émile Durkheim and his Sociology*, New York, 1939, p. [9].)

should best be presented.[1] In general, the translation holds as closely to the original as is feasible without being stiltedly literal. This principle may have resulted at places in a rather Gallic quality in the English, but this need not be altogether a bad thing.

Names of tribes and places are given in the orthography of the original ethnographic sources, which for the most part are in English in any case. Following ethnographic convention, tribal names do not receive an anglicized plural, so that for example 'les Zuñis' of the original becomes 'the Zuñi'. The footnotes have had to be renumbered. References have been abbreviated to author's name, year of publication, and page; and the complete particulars have been consolidated in a conventional bibliography. The sections in the original text have no headings, but brief indications of the areas they cover have been provided. An index has also been prepared.

These matters introduce the consideration of those respects in which this is an edition, and not simply an English rendering of a French text. The task was imposed by the surprising extent to which Durkheim and Mauss lapse from the conventional requirements of scholarly publication. Disregarding recognizable abbreviations of titles, and the simple omission of author's name, initials, or place of publication, there are no fewer than sixty-nine bibliographical errors, many of which definitely mislead a student seeking the sources of information used in the essay. To be particular, these comprise five instances in which the name of the author is misrendered (including one extreme case in which an article is attributed to a quite different person than the actual writer); twelve in which the title is substantially inaccurate; fifteen in which the year is incorrect;

---

[1] Cf. Translators' note to R. Hertz, *Death and The Right Hand*, 1960 (p. 5).

and thirty-seven wrong page-references. There are also more than a dozen mis-spellings of names of persons, places and things in the text. More seriously, there are a number of places at which Durkheim and Mauss misrender their sources, and at one point they cite a non-existent publication. As this rather dejecting catalogue implies, corrections have been made in all these cases: page-references have been checked, quotations are exactly repeated from the sources, and the items in the bibliography are complete in all the usual particulars. With these tacit changes, it is believed that the text is rid of any consequential inaccuracies or omissions.

However, it is assumed that ultimately, whatever the initial advantages, no one will rely exclusively on a translation for a scholarly purpose unless he has to; and it is expected that anyone making such a use of this edition will, if he can, refer to the original. It is to cope with this contingency in particular that attention is drawn, in bracketed footnotes, to certain of the more obvious discrepancies between the text and the translation.

Editing has been limited to checking Durkheim and Mauss's work in relation to the sources they used, and no attempt has been made to investigate any of the issues they raise by reference to other and subsequent publications in the very extensive literatures on Australia, North American Indians, or China. A great deal of the kind might be done, but such investigations would be of an ethnographic interest, whereas the present comments are concerned almost exclusively to assess Durkheim and Mauss's argument by reference to the facts which they employed and to principles which they might have acknowledged. Similarly, an examination of the precise connexion of the essay to the later writings of either Durkheim or Mauss belongs to their intellectual biographies, or to the history of science, and is

Introduction

not considered here.[1] It should be made clear, finally, that if their theoretical suggestions have not been developed, or other issues in the study of classification isolated, this is because such undertakings are more properly dealt with in original researches,[2] not in the restricted context of a critical introduction.

This edition was prepared with the aid of a Fellowship (1961–2) at the Center for Advanced Study in the Behavioral Sciences, Stanford, California. Grateful acknowledgement is made to the Center for the undistracted circumstances and the facilities which it provided, and to the University of Oxford for generously granting a dispensation from statutory duties which permitted them to be put to this use.

Thanks are due also to Professor Fred Eggan of the University of Chicago, who encouraged the project and made arrangement for publication, and to Dr. J. G. Vredenbregt of the University of Leiden for his help with bibliographic references in the Netherlands. Dr. V. W. Turner of Manchester University and Dr. D. H. P. Maybury-Lewis of Harvard University were kind enough to scan the Introduction and are warmly thanked for their friendly attentions.

R. N.

Merton College, Oxford

[1] There is one article, however, which deserves special mention with regard to the main contention of the essay as expressed in Durkheim's *Les Formes élémentaires de la Vie religieuse*, viz., P. M. Worsley, 'Émile Durkheim's theory of knowledge', *Sociological Review*, n.s., vol. 4, no. 1, 1956, pp. 47–62. Classification is particularly dealt with, on the basis of Dr. Worsley's own recent researches on Groote Eylandt, on pp. 54–62.

[2] Cf. Professor Lévi-Strauss's remarkable work *La Pensée Sauvage* (Paris, 1962), the central concern of which is primitive classification.

*On some primitive forms of classification :
contribution to the study of collective
representations*

THE PROBLEM

THE DISCOVERIES of contemporary psychology have thrown into prominence the frequent illusion that we regard certain mental operations as simple and elementary when they are really very complex. We now know what a multiplicity of elements make up the mechanism by virtue of which we construct, project, and localize in space our representations of the tangible world. But this operation of dissociation has been only very rarely applied as yet to operations which are properly speaking logical. The faculties of definition, deduction, and induction are generally considered as immediately given in the constitution of the individual understanding. Admittedly, it has been known for a long time that, in the course of history, men have learned to use these diverse functions better and better. But it is thought that there have been no important changes except in the way of employing them; that in their essential features they have been fully formed as long as mankind has existed. It has not even been imagined that they might have been formed by a painful combination of elements borrowed from extremely different sources, quite foreign to logic, and laboriously organized. And this conception of the matter was not at all surprising so long as the development of logical faculties was thought to belong simply to individual psychology, so long as no one had the idea of seeing in these methods of scientific thought veritable social institutions whose origin sociology alone can retrace and explain.

The preceding remarks apply particularly to what we

might call the classificatory function. Logicians and even psychologists commonly regard the procedure which consists in classifying things, events, and facts about the world into kinds and species, subsuming them one under the other, and determining their relations of inclusion or exclusion, as being simple, innate, or at least as instituted by the powers of the individual alone. Logicians consider the hierarchy of concepts as given in things and as directly expressible by the infinite chain of syllogisms. Psychologists think that the simple play of the association of ideas, and of the laws of contiguity and similarity between mental states, suffice to explain the binding together of images, their organization into concepts, and into concepts classed in relation to each other. It is true that recently a less simple theory of psychological development has come to the fore. The hypothesis has been put forward, namely, that ideas are grouped not only according to their mutual affinities but also according to the relations they bear to movements.[1] Nevertheless, whatever may be the superiority of this explanation, it still represents classification as a product of individual activity.

There is however one fact which in itself would suffice to indicate that this operation has other origins: it is that the way in which we understand it and practise it is relatively recent. For us, in fact, to classify things is to arrange them in groups which are distinct from each other, and are separated by clearly determined lines of demarcation. From the fact that modern evolutionism denies that there is an insuperable abyss between them, it does not follow that it so merges them as to claim the right to deduce one from the other. At the bottom of our conception of class there is the idea of a circumscription with fixed and definite outlines. Now one could almost say that this conception of classification does

[1] Münsterberg 1889/92, III, p. 113; II, 2nd fasc.; I, p. 129 etc.

not go back before Aristotle. Aristotle was the first to proclaim the existence and the reality of specific differences, to show that the means was cause, and that there was no direct passage from one genus to another. Plato had far less sense of this distinction and this hierarchical organization, since for him genera were in a way homogeneous and could be reduced to each other by dialectic.

Not only has our present notion of classification a history, but this history itself implies a considerable prehistory. It would be impossible to exaggerate, in fact, the state of indistinction from which the human mind developed. Even today a considerable part of our popular literature, our myths, and our religions is based on a fundamental confusion of all images and ideas. They are not separated from each other, as it were, with any clarity. Metamorphoses, the transmission of qualities, the substitution of persons, souls, and bodies, beliefs about the materialization of spirits and the spiritualization of material objects, are the elements of religious thought or of folklore. Now the very idea of such transmutations could not arise if things were represented by delimited and classified concepts. The Christian dogma of transubstantiation is a consequence of this state of mind and may serve to prove its generality.

However, this way of thinking exists today only as a survival, and even in this form it is found only in certain distinctly localized functions of collective thought. But there are innumerable societies whose entire natural history lies in etiological tales, all their speculation about vegetable and animal species in metamorphoses, all scientific conjecture in divinatory cycles, magical circles and squares. In China, in all the Far East, and in modern India, as well as in ancient Greece and Rome, ideas about sympathetic actions, symbolic correspondences, and astrological influences not only were or are very widespread, but exhausted or still exhaust collective knowledge. They all presuppose the belief in the possibility of the transformation of the most heterogeneous things one into another, and consequently the more or less complete absence of definite concepts.

If we descend to the least evolved societies known, those which the Germans call by the rather vague term *Naturvölker*, we shall find an even more general mental confusion.[1] Here, the individual himself loses his personality. There is a complete lack of distinction between him and his exterior soul or his totem. He and his 'fellow-animal' together compose a single personality.[2] The identification is such that the man assumes the characteristics of the thing or animal with which he is thus united. For example, on Mabuiag Island people of the crocodile clan are thought to have the temperament of the crocodile: they are proud, cruel, always ready for battle.[3] Among certain Sioux, there is a section of the tribe which is called red, and which comprises the clans of the mountain lion, buffalo, and elk, all animals characterized by their violent instincts; the members of these clans are from birth warriors, whereas the farmers, people who are naturally peaceful, belong to clans of which the totems are essentially pacific animals.[4]

If it is thus with people, all the more reason that it should be the same with things. Not only is there complete indifferentiation between sign and thing, name and person, places and inhabitants, but, to adopt a very exact remark made by von den Steinen concerning the Bakairi[5] and the Bororo, for the primitive the principle of *generatio aequivoca* is proved.[6] The Bororo sincerely imagines himself to be a parrot; at least, though he assumes the characteristic form

---

[1] Cf. Bastian 1887, pp. 11 and 83; 1886, vol. I, p. 18.
[2] Spencer and Gillen 1899, pp. 107 and 207.
[3] Haddon 1901, p. 132.   [4] Dorsey 1884, p. 208.
[5] Former Caribs, at present situated on the Xingu.
[6] Von den Steinen 1894, p. 352.

only after he is dead, in this life he is to that animal what the caterpillar is to the butterfly. The Trumaí are genuinely thought to be aquatic animals. 'The Indian lacks our determination of genus, such that one does not mix with the other.' [1] Animals, people, and inanimate objects were originally almost always conceived as standing in relations of the most perfect identity to each other. The relations between the black cow and rain, between the white or red horse and the sun, are characteristic traits of the Indo-European tradition;[2] and examples could be multiplied infinitely.

Besides, this state of mind does not differ appreciably from that which still, in each generation, serves as point of departure for the development of the individual. Consciousness at this point is only a continuous flow of representations which are lost one in another, and when distinctions begin to appear they are quite fragmentary. This is to the right, that to the left; that is past, this is present; this resembles that, this accompanies that. This is about all that even the adult mind could produce if education did not inculcate ways of thinking which it could never have established by its own efforts and which are the result of an entire historical development. It is obvious what a great difference there is between these rudimentary distinctions and groupings and what truly constitutes a classification.

Far, then, from man classifying spontaneously and by a sort of natural necessity, humanity in the beginning lacks the most indispensable conditions for the classificatory function. Further, it is enough to examine the very idea of classification to understand that man could not have found its essential elements in himself. A class is a group of things; and things do not present themselves to observation

---

[1] Von den Steinen 1894, p. 351.
[2] Caland 1901; Hillebrandt 1897, p. 120; von Negelein 1901.

7

grouped in such a way. We may well perceive, more
or less vaguely, their resemblances. But the simple fact of
these resemblances is not enough to explain how we are led
to group things which thus resemble each other, to bring
them together in a sort of ideal sphere, enclosed by definite
limits, which we call a class, a species, etc. We have no
justification for supposing that our mind bears within it at
birth, completely formed, the prototype of this elementary
framework of all classification. Certainly, the word can help
us to give a greater unity and consistency to the assemblage
thus formed; but though the word is a means of realizing
this grouping the better once its possibility has been conceived, it could not by itself suggest the idea of it. From
another angle, to classify is not only to form groups; it
means arranging these groups according to particular relations. We imagine them as co-ordinated, or subordinate one
to the other, we say that some (the species) are included in
others (the genera), that the former are subsumed under the
latter. There are some which are dominant, others which
are dominated, still others which are independent of each
other. Every classification implies a hierarchical order for
which neither the tangible world nor our mind gives us the
model. We therefore have reason to ask where it was
found. The very terms which we use in order to characterize
it allow us to presume that all these logical notions have an
extra-logical origin. We say that species of the same genera
are connected by relations of kinship; we call certain classes
'families'; did not the very word genus (*genre*) itself originally designate a group of relatives (*γένος*)? These facts
lead us to the conjecture that the scheme of classification is not
the spontaneous product of abstract understanding, but results
from a process into which all sorts of foreign elements enter.

Naturally, these preliminary observations are in no way
intended to resolve the problem, or even to prejudge its

solution, but merely to show that there is a problem which must be posed. Far from being able to say that men classify quite naturally, by a sort of necessity of their individual understandings, we must on the contrary ask ourselves what could have led them to arrange their ideas in this way, and where they could have found the plan of this remarkable disposition. We cannot even dream of tackling this question in all its ramifications. But, having posed it, we should like to adduce certain evidences which, we believe, may elucidate it. The only way to answer it is to investigate the most rudimentary classifications made by mankind, in order to see with what elements they have been constructed. So in what follows we shall report a number of classifications which are certainly very primitive and the general meaning of which seems not to be in doubt.

This question has not yet been put in the terms that we have just enunciated. But among the facts which we shall use in this work there are some which have already been noticed and studied by a number of authors. Bastian has on a number of occasions occupied himself with cosmological notions in general, and has quite often attempted to systematize them.[1] But he has concerned himself mostly with cosmologies of oriental peoples and with those of the Middle Ages, and has reported the facts rather than sought to explain them. As for more rudimentary classifications, first Howitt[2] and then Frazer[3] have already given a number of examples. But neither has seen their importance from the point of view of the history of logic. As we shall see, indeed, Frazer's interpretation of the facts is exactly the opposite of that which we shall propose.

[1] 1887, with an interesting atlas; 1892; etc.
[2] Fison and Howitt 1880, p. 168; Howitt 1889a, p. 61. Howitt says textually: 'This is not peculiar to these tribes but is found at far distant places in Australia, and may be much more general than has been suspected.'   [3] Frazer 1887, p. 85; 1899.

Chapter One

THE AUSTRALIAN TYPE OF CLASSIFICATION

THE MOST SIMPLE SYSTEMS of classification known are those found among the tribes of Australia.

The most widespread form of social organization among these societies is well known. Each tribe is divided into two large fundamental sections which we shall call moieties.[1] Each moiety, in turn, comprises a certain number of clans, i.e. groups of individuals with the same totem. In principle, the totems of one moiety are not found in the other. In addition to this division into clans, each moiety is divided into two classes which we shall call 'marriage classes'. We give them this name because their purpose, above all, is to regulate marriage: a particular class of one moiety may marry only

[1] This terminology is not adopted by all authors. Many prefer to use the word 'classes'. This leads to regrettable confusions with marriage classes, which are dealt with below. In order to avoid these errors, whenever an author calls a moiety a class we shall replace the latter word by the former. Unity of terminology will facilitate the comprehension and the comparison of the facts. It would be a good thing, in any case, if the meanings of a terminology so often employed could be agreed upon. [Durkheim and Mauss use the term 'phratry' (*phratrie*), but 'moiety' has long been the means of the terminological agreement they wished for. The term 'phratry' commonly means one of a number, more than two, of groupings of clans; and this indeed is one sense in which Durkheim and Mauss themselves also use it, as when they describe Loucheux society (below, p. 63, n. 4) as comprising three phratries.—R. N.]

10

with a particular class of the other moiety. The over-all organization of the tribe thus has the following form.[1]

moiety I $\begin{cases} \text{marriage class A} \\ \text{marriage class B} \end{cases}$ $\left.\begin{array}{l} \\ \\ \end{array}\right\}$ emu clan
snake clan
caterpillar clan, etc.

moiety II $\begin{cases} \text{marriage class A}' \\ \text{marriage class B}' \end{cases}$ $\left.\begin{array}{l} \\ \\ \end{array}\right\}$ kangaroo clan
opossum clan
crow clan, etc.

The classes designated by the same letter, A, A' and B, B', are those which maintain connubium.

All the members of the tribe are classed in this way in definite categories which are enclosed one in the other. *Now the classification of things reproduces this classification of men.*

Cameron has already observed that among the Ta-ta-thi 'everything in the universe is divided among the different members of the tribe'. 'Some', he says, 'claim the trees, others the plains, others the sky, stars, wind, rain, and so forth.'[2] Unfortunately, this information lacks precision. We are not told to which groups of individuals the different groups of things are related in this way.[3] But we have facts from another source which are extremely significant.

The tribes of the Bellinger River are each divided into

[1] This scheme represents only the organization which we consider typical. It is the most general. But in certain cases it is only found in an altered form. In one place, the totemic classes have clans and are replaced by purely local groups; in another, neither moieties nor classes are to be found.—To be quite complete, we should even have to add a division into local groups which is often superimposed on the above divisions.

[2] Cameron 1885, p. 350. It is not explicitly said, moreover, that this relates only to the Ta-ta-thi. The preceding paragraph mentions a whole group of tribes.

[3] It seems, however, that it is a question of a division into totemic groups, analogous to that which will be discussed below. But this is only a supposition.

two moieties; and, according to Palmer, this division applies equally to nature. 'All nature is divided into class names[1] and said to be male and female. The sun and moon and stars are said to be men and women, and to belong to classes just as the blacks themselves.' [2] This tribe is fairly close to another tribe, that of Port Mackay in Queensland, in which we find the same system of classification. According to the answers made by Bridgeman to the questionnaires of Curr, Smyth, and Lorimer Fison, this tribe, like its neighbours, is divided into two moieties, one called Youngaroo, the other Wutaroo. As a matter of fact, there are marriage classes as well; but these do not appear to have affected cosmological notions. On the contrary, the division into moieties is considered 'as a universal law of nature'. 'All things, animate and inanimate,' says Curr after Bridgeman, 'are divided by these tribes into two classes, named *Youngaroo* and *Wootaroo*.' [3] The same observer reports (according to Smyth) that 'they divide everything into moieties. They tell you that alligators are Youngaroo and kangaroos are Wootaroo—the sun is Youngaroo and the moon is Wootaroo; and so on with the constellations, with the trees, and with the plants.' [4] And Fison relates that: 'Everything in nature, according to them, is divided between the two classes. The wind belongs to one, and the rain to the other. . . . If a star is pointed out they will tell you to which division [moiety] it belongs.' [5]

Such a classification is of extreme simplicity, since it is simply bipartite. Everything is distributed in the two cate-

---

[1] [In this footnote Durkheim and Mauss report that they render the original words 'class' by their term 'phratry', and advise the reader that henceforth they will make the substitution without warning. In this edition the original wording of all quotations is adhered to.—R. N.]

[2] Palmer 1884, p. 300; cf. p. 248.

[3] Curr 1886–7, vol. III, p. 45.

[4] Smyth 1878, vol. I, p. 91.   [5] Fison and Howitt 1880, p. 168.

gories corresponding to the two moieties. The system becomes more complex when it is no longer only the division
into moieties which is the framework for the division of
things, but also the division into four marriage classes. This
is the case among the Wakelbura of north-central Queensland. Muirhead, a settler who lived a long time in the area
and was an acute observer, sent information on a number of
occasions to Curr and to Howitt concerning the organization of these peoples and their cosmology, and these reports,
which appear to apply to a number of tribes,[1] have been
corroborated by another observer, Lowe.[2] The Wakelbura
are divided into moieties, Mallera and Wutaru: each is
further divided into two marriage classes. The classes of the
Mallera moiety bear the names Kurgila and Banbey; those
of the Wutaru moiety, Wongu and Obù. Now these two
moieties and the marriage classes* 'divide the whole universe into groups'. Howitt writes that 'The two primary
classes are Mallera and Wutheru [equivalent to Wutaru];
therefore, all objects are either one or the other.' [3] Similarly,
according to Curr, food eaten by the Banbey and the Kargilla is called Mullera, and that of the Wongoo or Oboo
(Obù) is called Woothera (Wutaru).[4] But we find in addition a distribution by marriage classes. 'Certain classes are
allowed to eat only certain sorts of food. Thus, the Banbey
are restricted to opossum, kangaroo, dog, honey of small bee,
etc. To the Wongoo is allotted emu, bandicoot, black duck,
black snake, brown snake, etc. The Oboo rejoice in carpet
snakes, honey of the stinging bees, etc. etc. The Kargilla

[1] Howitt 1889a, p. 61, fn. 3.   [2] Curr 1886–7, vol. III, p. 27.

* [The text has 'two marriage classes'.—R. N.]

[3] Howitt 1889b, p. 326; 1889a, p. 61, fn. 3 [authors' italics].

[4] Curr 1886–7, vol. III, p. 27. We have corrected Curr's statement,
which, due evidently to a misprint, says that food eaten by the Wongu is
called Obu or Wuthera. It would have been better in any case to write
Wongoo *and* Oboo. [Durkheim and Mauss themselves write 'Obu *and*
Wuthera'.—R. N.]

13

live on porcupine, plain turkey, etc. etc. and to them, it appears, belong water, rain, fire, and thunder. . . . There are innumerable articles of food, fish, flesh and fowl, into the distribution of which Mr. Muirhead does not enter.' [1]

To be exact, there does seem to be some uncertainty in the reports on this tribe. According to what Howitt says, one might believe that the division is made by moieties and not by marriage classes. Thus the things attributed to the Banbey and the Kargilla would all be Mallera. [2] But the divergence is only in appearance, and is indeed instructive. In fact the moiety is the genus, the marriage class is the species; the name of the genus applies to the species, which is not to say that the species has not its own. Just as the cat falls under the class of quadrupeds and may be designated by this name, so things belonging to the Kargilla species belong to the superior genus Mallera (moiety) and may consequently be themselves called Mallera. This proves that we are no longer dealing with a simple dichotomy of things into opposed kinds, but with hierarchized concepts included within each of these kinds.

The importance of this classification is such that it extends to all the facts of life; its impress is seen in all the principal

[1] Curr 1886–7, vol. III, p. 27. It will be noticed that each moiety or class seems to eat the flesh of animals which are thus assigned to it. Now, as we shall have occasion to discuss below, the animals thus attributed to a moiety or class are generally of a totemic character, and consequently to eat them is forbidden to the groups of individuals to which they are assigned. Does the contrary fact reported from the Wakelbura constitute a case of the ritual consumption of the totemic animal for the corresponding totemic group? We cannot say. Perhaps, too, there is some error of interpretation in this observation, a mistake always easy to make in matters as complex and difficult to apprehend as these. It is indeed rather remarkable that, according to the table given to us, the totems of the Mallera moiety are the opossum, bush-turkey, kangaroo, and the small bee, all creatures whose consumption is permitted precisely to two marriage classes of this moiety, viz. the Kurgil and the Banbey (cf. Howitt 1883, p. 45; 1884b, p. 337).

[2] Howitt 1884c, p. 438, fn. 2.

rites. Thus, a sorcerer belonging to the Mallera phratry may use in his art only things which similarly belong to Mallera.[1] At a burial, the scaffold on which the corpse is exposed (assuming still that we are concerned with a Mallera) 'must be made of the wood of some tree belonging to the Mallera class'.[2] The same applies to the branches which cover the corpse. If it is a Banbey, a broad-leafed box tree must be used, for this tree is Banbey;[3] and it is men of the same phratry who will carry out the rite. The same organization of ideas is the basis of precognition; it is by taking them as premiss that dreams are interpreted,[4] that causes are determined, and responsibilities assigned. It is well known that in this kind of society death is never considered as a natural event, due to the action of purely physical causes; it is almost always attributed to the magical influence of some sorcerer, and the determination of the guilty party forms an integral part of the funerary rites. Now among the Wakelbura it is the classification of things by moiety and marriage class which furnishes the means of discovering the class to which the person responsible belongs, and perhaps the very individual.[5] The warriors smooth out the earth under the scaffold on which the corpse rests, and round about it, so that the slightest mark shall be visible. The next day they carefully examine the ground under the corpse. If an animal has passed by, its tracks are easily discovered; from these the blacks infer the class of the person who has caused the death of their relative.[6] For example, if the tracks of a wild dog are found they will know that the murderer is a Mallera and a

---

[1] Howitt 1889b, p. 326; 1889a, p. 61, fn. 3.

[2] Howitt 1889b, p. 326; cf. 1889a, p. 61, fn. 3.

[3] Howitt 1884a, p. 191, fn. 1.

[4] Curr 1886–7, vol. III, p. 27. 'Should a Wongoo Black, camped out by himself, dream that he has killed a porcupine, he would believe that he would see a Kargilla Black next day.'

[5] Howitt 1884a, p. 191, fn. 1.   [6] Curr 1886–7, vol. III, p. 28.

Banbey; for this animal belongs to this moiety and to this marriage class.[1]

There is yet more to the matter than this. This logical order is so rigid, the power of constraint of these categories on the mind of the Australian is so strong, that in certain cases a whole group of acts, signs, and things may be seen to be arranged according to these principles. When an initiation ceremony is to be held, the local group which takes the initiative in calling together the other local groups belonging to the same totemic clan gives them warning of it by sending a 'message stick' which must belong to the same moiety as the senders and the bearer.[2] This obligatory concordance may not seem at all extraordinary, seeing that almost everywhere in Australia the invitation to an initiation is delivered by a messenger carrying 'devils' (or bull-roarer, *turndun*, *churinga*) which are evidently the property of the whole clan, and consequently of the host group as well as of the guest group.[3] But the same rule applies to messages intended to effect a hunting rendezvous, and in this case the sender, the recipient, the messenger, the wood of the message-stick, the game designated, the colour with which it is painted, everything rigorously agrees according to the principle set by the classification.[4] Thus, in an example reported by Howitt,[5] the stick was sent by an Obù. Consequently, the wood of the stick was of *gidyea*, a sort of acacia

[1] Curr even seems to mention, in this connexion, that these animals are indeed totems; and that they are the same as the prescribed foods: 'the murder [would] be assigned to some member of the tribe in whose dietary scale the animal, bird, or reptile is included. If a carpet snake, an Obad; . . . Then would come the consideration, to what particular . . . Obad suspicion should attach.'

[2] Howitt 1884c, p. 438, fn. 2; cf. Howitt 1889b, Pl. XIV, Fig. 13.

[3] See examples in Howitt 1884c, p. 438.

[4] Howitt 1889b, p. 326; Pl. XIV, Figs. 15 and 16. [The authors refer to Figures '25, 16, 136'; but there are only 17 figures on the Plate. At the place cited, Howitt refers to Figs. 15 and 16 only.—R. N.]

[5] Howitt 1889b, p. 326.

belonging to the Wutaru moiety of which the Obù are part. The game represented on the stick was the emu and the wallaby, animals of the same moiety. The colour of the stick was blue, probably for the same reason. Thus everything follows, as in a theorem: the sender, the recipient, the object and the writing of the message, the wood employed, everything is related. All these ideas seem to the primitive to be subject to a logical necessity by which they are entailed.[1]

Another system of classification, more complete and perhaps more characteristic, is that in which things are no longer distributed by moiety and marriage class, but by moieties and by clans or totems. 'Australian totems', says Fison, 'have a special value of their own. Some divide, not mankind only, but the whole universe, into what may almost be called gentile divisions.'[2] There is a very simple reason for this. It is that if totemism is, in one aspect, the grouping of men into clans according to natural objects (the associated totemic species), it is also, inversely, a grouping of

---

[1] Muirhead says explicitly that this procedure is followed by the neighbouring tribes.—It may be justifiable to relate to the Wakelbura system the facts reported by Roth concerning the Pitta-Pitta, Kalkadoon, Mitakoodi, and the Woonamurra, all neighbours of the Wakelbura (Roth 1897, pp. 47, 48; cf. *Proceedings of the Royal Society of Queensland*, 1897). Each marriage class has a series of dietary prohibitions of such a kind that 'all the food at the disposal of the tribe is divided among its members'. Let us take as example the Pitta-Pitta. Individuals of the Koopooroo class may not eat iguana, yellow dingo, small fish 'with a bone in it' (p. 57). The Wongko have to avoid bush turkey, bandicoot, eagle-hawk, black dingo, 'absolutely white' duck, etc.; the Koorkilla are forbidden the kangaroo, carpet snake, carp, a duck with brown head and large belly, different species of diving birds, etc.; the Bunburi are forbidden the emu, yellow snake, a certain kind of hawk, and a certain kind of parakeet. We have here, in any case, an example of classification which extends at least to a particular group of objects, viz. products of the hunt. And this classification is modelled on that of the tribe into four marriage classes or 'paedo-matronymic' groups, as our author puts it. Roth does not appear to have investigated whether this division is extended to the rest of things in nature.

[2] Fison and Howitt 1880, p. 167.

natural objects in accordance with social groups. The same observer continues: 'The Southern Australian savage looks upon the universe as the Great Tribe to one of whose divisions he belongs; and all things, animate and inanimate, which belong to his class are parts of the body corporate whereof he himself is a part. They are "almost parts of himself", as Mr. Stewart shrewdly remarks.' [1]

The best known example of these facts is that to which Fison, Smyth, Curr, Andrew Lang, and Frazer have successively drawn attention.[2] It concerns the Mount Gambier tribe. The information comes from Stewart, who knew this tribe intimately. It is divided into moieties, one of which is called Kumite, the other Kroki: both these names, moreover, are widespread in the whole of south Australia, where they are used with the same meaning. Each of these moieties is itself divided into five matrilineal totemic clans.[3] It is among these clans that things are divided. None of these clans may eat any of the comestible objects which are thus attributed to it. 'A man does not kill, or use as food, any of the animals of the same sub-division with himself.' [4] But, beside these forbidden animals and even vegetables,[5] an indefinite multitude of things of all sorts is attached to each class.

'The Kumite and Krokee [Kroki] classes are each divided into five sub-classes [*sc.* totemic clans], under which are ranked certain objects which they call *tooman* = *flesh* or *wingo* = *friend*. All things in nature belong to one or other

[1] Fison and Howitt 1880, p. 170. Cf. Smyth 1878, vol. I, p. 92, who understands and underlines the importance of this fact, about which he says 'there is a great deal yet to be ascertained'.

[2] Smyth 1878, vol. I, p. 92; Fison and Howitt 1880, p. 168; Lang 1896, p. 132; Frazer 1887, p. 85; Curr 1886–7, vol. III, p. 462. Our account is based on Curr and on Fison and Howitt.

[3] Curr 1886–7, vol. III, p. 461.

[4] Stewart, in Fison and Howitt 1880, p. 169.

[5] Curr 1886–7, vol. III, p. 462.

of these ten sub-classes.' [1] Curr indicates, but only by way of examples, certain of the things which are classed in this way.

The first[2] of the Kumite totems[3] is that of the Mula or fishhawk; to it belong—or, as Fison and Howitt put it, in it are included—smoke, honeysuckle, trees, etc.[4]

The second is that of the Parangal or pelican, to which belong a tree with black wood, dogs, fire, ice, etc.

The third is that of the Wa or crow, under which are subsumed rain, thunder, lightning, hail, clouds, etc.

The fourth totem is that of the Wila or black cockatoo, to which are related the moon, stars, etc.

Lastly, to the totem of the Karato (harmless snake) belong the fish, stringybark tree, salmon, seal, etc.[5]

We have less information on the totems of the Kroki phratry. We know only three of them. With the Werio (tea-shrub) totem are connected ducks, wallabies, hens, crayfish, etc.; with that of the Murna (a sort of edible root),[6] the buzzard, dolvich (sort of a small kangaroo), quails, etc.; with that of the Karaal (white crestless cockatoo),[7] kangaroo, mock oak, summer, the sun, autumn (feminine gender), and the wind (same gender).

[1] Curr 1886–7, vol. III, p. 461.

[2] Curr says expressly that they are only examples.

[3] This expression should not be taken to imply that there is a hierarchy of clans. The order in which they are given by Fison is not the same as that given by Curr. We follow Fison.

[4] The name of each totem is preceded by the prefix Burt or Boort, meaning 'dry'. We omit it from the list.

[5] This 'etc.' indicates that the list of things subsumed under the totem is not exhaustive.

[6] According to Curr, the totem is that of the turkey (*laa*) and includes certain edible roots among the things connected with it. There is nothing surprising in these variations. They merely prove that it is often difficult to determine exactly which thing it is, among those which are classed under the clan, that serves as totem for the whole group.

[7] Fison says that the totem is the black cockatoo. This is undoubtedly a mistake. Curr, who simply copies the information of Stewart, says it is white, which is very likely the case.

The Australian type of classification

We are thus in the presence of a still more complex and extensive system than the preceding ones. It is no longer a question simply of a classification into two fundamental genera (moieties), each comprising two species (the two marriage classes). Certainly, the number of fundamental genera is the same here, too, but that of the species of each genus is much more considerable, since the clans may be very numerous. But, at the same time, the state of initial confusion from which the human mind has developed is still perceptible in this more differentiated organization. Though the distinct groups are multiplied, within each elementary group the same indistinction reigns. Things attributed to one moiety are clearly separated from those which are attributed to the other; those attributed to different clans of one and the same moiety are no less distinct. But all those which are included in one and the same clan are, in large measure, undifferentiated. They are of the same nature; there are no sharp lines of demarcation between them such as exist between the ultimate varieties of our classifications. The individuals of the clan, the creatures of the totemic species, and those of related species, all these are nothing but diverse aspects of one and the same reality. The social divisions applied to the primitive mass of representations have indeed cut them into a certain number of delimited divisions, but the interior of these divisions has remained in a relatively amorphous state which testifies to the slowness and the difficulty with which the classificatory function has been established.

In some cases it is perhaps not impossible to perceive certain of the principles according to which these groups are constituted. Thus, in this Mount Gambier tribe the sun, summer, and the wind are connected with the white cockatoo; the moon, stars and falling stars are linked to the black cockatoo. It seems that colour has provided the line according to which these diverse representations are antithetically arranged. Similarly, the crow quite naturally, by virtue of its colour, covers the rain and consequently winter, clouds, and—through these—lightning and thunder. When Stewart asked a native to which division the bull belonged, he received, after a moment of reflection, the following answer: 'It eats grass: it is Boortwerio,' i.e. of the tea-shrub clan, which probably comprises all grasslands and herbivores.[1] But this is very probably an *ad hoc* explanation to which the black has recourse in order to justify his classification to himself and to reduce it to general rules by which to be guided. Quite often, moreover, such questions take him unawares, and he is constrained, in answer to everything, to invoke tradition.

The reasons which have led to the establishment of the categories have been forgotten, but the category persists and is applied, well or ill, to new ideas such as that of the bullock which has only very recently been introduced.[2] All the more reason for us not to be surprised that many of these associations pass us by. They are not products of a logic identical with our own. They are governed by laws which we do not suspect.

A similar case is provided by the Wotjobaluk, a tribe in New South Wales, and one of the most evolved of all the Australian tribes. We owe our information to Howitt himself, whose competence is well known.[3] The tribe is divided into two moieties, Krokitch and Gamutch,[4] which, he says,

---

[1] Fison and Howitt 1880, p. 169. [Durkheim and Mauss write 'therefore it is Boortwerio', thus making explicit a deduction on the part of the native which may only be inferred; and they include the remainder of the sentence in the quotation marks as though the further explanation were that of the informant himself.—R. N.]

[2] [Durkheim and Mauss place this sentence within quotation marks and ascribe it to Fison and Howitt 1880, p. 169; but it exists nowhere in that source.—R. N.]　　　　[3] Howitt 1889a, pp. 60 ff.

[4] The kinship of these names to the Kroki and Kumite of the Mount

seem in fact to divide all natural objects between them. As the natives say, 'they belong to them'. Further, each moiety comprises a certain number of clans. By way of example, Howitt cites the clans of hot wind, white crestless cockatoo, and things belonging to the sun, in the Krokitch moiety; and, in the Gamutch moiety, those of the deaf adder, black cockatoo, and pelican.[1] But these are only examples; he says, 'I have given three totems of each class as examples, but there are more; of Krokitch, eight, and of Gamutch, at least four.'[2] Now things classed in each moiety are divided among the different clans of which it is composed. In the same way as the primary division (or moiety) is split into a number of totemic divisions, similarly all the objects attributed to the moiety are divided among these totems. Each totem thus possesses a certain number of natural objects, not all of which are animals, since among them there are a star, fire, wind, etc.[3] Things thus classed under each totem are called by Howitt sub-totems or pseudo-totems. The white cockatoo, for example, includes fifteen and the hot wind five.[4] Finally, the classification is pushed to such a degree of complexity that sometimes tertiary totems are found subordinated to the secondary. Thus the Gamutch class (moiety) includes the pelican division (totem); the pelican comprises further sub-divisions (sub-totems, species of things classed under the totem) among which is fire; and fire itself includes signals (probably made with the aid of fire) as a tertiary sub-division.[5]

[1] Howitt 1885, p. 818.
[2] Howitt 1885, p. 818; 1889a, p. 61.
[3] Howitt 1889a, p. 61.
[4] Howitt 1885, p. 818.
[5] The term by which the individuals composing this sub-division of the sub-clan call themselves means exactly: 'we are warming ourselves'

Gambier tribe is clear; which proves the authenticity of this system of classification, which is thus found at points quite distant from each other.

The Australian type of classification

This curious organization of ideas, parallel to that of the society, is perfectly analogous, except for its complication, to that which we have found among the Mount Gambier tribes; it is equally analogous to the division by marriage classes which we have observed in Queensland, and to the dichotomous division by moieties which we have found practically everywhere.[1] But having described the different varieties of this system, such as they function in these societies, in an objective fashion, it would be interesting to know how the Australian himself sees them; what idea he himself conceives of the relations between the groups of things thus classed. In this way we could realize better what the logical ideas of primitive man are and the way in which they are formed. Now we do have information, concerning the Wotjobaluk, which permits us to clear up certain matters of this kind.

As we might have expected, this representation appears under different aspects.

First of all, these logical relations are conceived in the form of more or less close kinship relations with respect to

[1] We leave on one side the influence which the division of individuals into clearly differentiated sexual groups could have had on the division of things into genera. Nevertheless, wherever each sex has its own totems it is difficult to believe that this influence should not have been considerable. We confine ourselves to indicating the problem as examined by Frazer (see *Année Sociologique*, vol. IV, 1901, pp. 364–5).

(Howitt 1889a, p. 61). [Durkheim and Mauss misread the English, which they render as *nous avertissons*, 'we are warning'. Howitt says the name was given 'because fire . . . is one of their pseudo-totems' (p. 61). There is nothing about signals.—R. N.] To have an exact idea of the complexity of this classification, one more element should be added. Things are not only distributed among the clans of the living, but the dead also form clans which have their own totems and consequently their own things which are attributed to them. These are what are called mortuary totems. Thus when a Krokitch of the Ngaui (sun) totem dies, he loses his name and ceases to be Ngaui, and becomes Mitbagragr, bark of the Mallee tree (Howitt 1889a, p. 64). On the other hand, there is a relation of dependence between the totems of the living and those of the dead. They belong to the same system of classification.

23

the individual. When the classification is made simply by moieties, without any further sub-division, everyone regards himself as a relative, and equally a relative, of the beings attributed to the moiety of which he is a member; they are all, by the same title, his flesh, his friends, whereas he has quite other feelings about the beings of the other moiety. But when a division into classes or clans is superimposed on to this fundamental division, these kinship relations are differentiated. Thus a Kumite of Mount Gambier feels that everything Kumite is his; but some are closer in that they are of his totem. The kinship, in this latter case, is more close. 'The class name is general,' says Howitt; 'the totem name is, in one sense, individual, for it is certainly nearer to the individual than the name of the moiety of the community to which he belongs.' [1] Things are thus conceived as disposed in a series of concentric circles around the individual; the more distant, those which correspond to the widest genera, are those comprising things which touch him least; they become progressively differentiated as they close in upon him. Thus, when they are foodstuffs, it is only the closest which are forbidden to him.[2]

In other cases, these relations are thought of as relations between possessors and things possessed. The difference between totems and sub-totems, according to Howitt, is as follows: 'Both are called "mir", but while one of my informants, a Krokitch man, *takes* his name Ngaui from the sun [totem properly speaking], he *owns* Bunjil, one of the fixed stars [which is a sub-totem]. . . . The true totem owns him, but he owns the pseudo-totem.' [3] Similarly, a member of the Wartwut (hot wind) clan 'specially claimed as "belonging" to him' [4] one of the five sub-totems, viz. Moiwuk

[1] Howitt 1885, p. 819.

[2] See above, p. 19, n. 6, concerning the Mount Gambier tribe.

[3] Howitt 1889a, pp. 61–2, 64 [italics supplied by Durkheim and Mauss].   [4] Howitt 1885, p. 819.

(carpet snake). To put it precisely, it is not the individual who in himself possesses the sub-totem: it is to the principal totem that those who are subordinated to it belong. The individual is only an intermediary in this situation. It is because he has the totem in himself (as equally do all the members of the clan) that he has a sort of proprietary right in the things attributed to this totem. Moreover, behind the statements which we have just quoted one senses also something of the conception which we first set ourselves to analyse. For a thing 'which belongs especially to an individual' is also closer to him and concerns him more particularly.[1]

It is true that in certain cases the Australian seems to conceive the hierarchy of things in an exactly inverse order. It is then the most distant which he considers the most important. One native, of whom we have already spoken, who had the sun (Ngaui) as totem and a star (Bunjil) as subtotem, said 'that he is Ngaui, but not Bunjil'.[2] Another whom we have also mentioned, whose totem was Wartwut (hot wind) and sub-totem Moiwuk (carpet snake), was, as he was even advised by one of his companions, 'Wartwut but also *partly* Moiwuk'.[3] Only a part of him is carpet snake. This is what is meant by another statement reported by Howitt. A Wotjobaluk often has two names, one being his totem and the other his sub-totem. The former is really his name, the other 'comes a little behind it';[4] it is secondary

[1] The preceding texts concern only the relations of sub-totem to totem, not those of the totem to the moiety. But, clearly, the latter must have been conceived in the same manner. If we have no texts giving us information especially on this point, this is because the moiety no longer plays much part in these tribes and has a lesser place in their preoccupations.

[2] See above, p. 24.

[3] Howitt 1889a, p. 63 [italics supplied by the authors]. In the text the name is given as Moiwiluk; it is a synonym of Moiwuk.

[4] Howitt 1889a, p. 61.

in rank. This means that the things which are most essential to an individual are not those which are closest to him, those which have most to do with his individual personality. The essence of man is humanity. The essence of the Australian is in his totem, and even in the collection of things which characterize his moiety, rather than in his sub-totem. There is thus nothing in these accounts which contradicts the preceding ones. The classification continues to be conceived in the same manner, except that its constituent relations are considered from another point of view.

Chapter Two

OTHER AUSTRALIAN SYSTEMS

HAVING ESTABLISHED this type of classification, we have now to try, as far as possible, to determine its generality.

The facts do not permit us to say that it is found everywhere in Australia, nor that it has the same distribution as a tribal organization into moieties, marriage classes, and totemic clans. We believe that it would doubtless be found, whether complete or in altered form, if it were looked for in numbers of Australian societies in which it has not yet been noticed; but we may not prejudge the result of observations which have not been made. Nevertheless, the sources which we already possess allow us to be sure that it certainly is, or has been, very widespread.

First of all, in many cases where our form of classification has not been directly observed, secondary totems have however been found and reported which, as we have seen, presuppose it. This is particularly true of the islands of the Torres Straits, near New Guinea. On Kiwai nearly all the clans have vegetable species for totems (*miramara*); one of these, palm tree (*nipa*), has as secondary totem the crab which lives in the tree of the same name.[1] On Mabuiag (an island to the west of the Torres Straits)[2] we find an

[1] Haddon 1901, p. 102.
[2] Since the reports of Haddon (1901, p. 102; 1890, p. 39) we know

27

organization of clans into two moieties: that of the little *augŭd* (totem), and that of the great *augŭd*. One is the land moiety, the other the sea moiety; one camps downwind, the other upwind; one to the south-east, the other to the north-west. The totems of the sea moiety are the dugong and a creature which Haddon calls the shovel-nose skate; the totems of the other, with the exception of the crocodile, which is amphibious, are all terrestial: viz. crocodile, snake, cassowary.[1] These are obviously important indications of the classification. But, even more, Haddon expressly mentions 'secondary, or properly speaking subsidiary, totems': the hammer-headed shark, the shark, tortoise, and sting ray belong, as such, to the sea moiety; the dog, to that of the land. Two other sub-totems, in addition, are attributed to the latter: these are crescent-shaped ornaments made of turtle-shell.[2] Keeping in mind that totemism everywhere in these islands is in full decline, it seems the more legitimate to see in these facts the relics of a more complete system of classification.—It is quite possible that a similar organization is to be found elsewhere in the Torres Straits and in the interior of New Guinea. The fundamental principle, that of the division by moieties, and clans grouped three by three, has been clearly reported from Saibai (an island in the strait) and in Daudai.[3]

We should be tempted to discern traces of this same classification on the islands of Murray, Mer, Waier, and Dauar.[4] Without going into the details of the social organization, as described by Hunt, we should like to draw attention to the following fact. A number of totems exist among these

[1] Haddon 1901, p. 132. But the names that we give to the moieties are not given by Haddon.
[2] Haddon 1901, p. 138; cf. Rivers 1900, pp. 75 ff.
[3] Haddon 1901, p. 171.   [4] Hunt 1899, pp. 5 ff.

that totemism is found only in the western islands and not in those of the east.

peoples. Now each one of these confers upon the individuals who belong to it various powers over different kinds of things. Thus, people who have the drum as totem possess the following powers: they have the right to conduct a ceremony which consists in imitating dogs and beating drums; they supply the magicians who have to secure the multiplication of tortoises, assure the banana crop, and divine the identities of murderers from the movements of a lizard; and, finally, it is they who impose the snake taboo. It is thus possible to say with fair likelihood that to the drum clan belong, in certain respects and besides the drum itself, the snake, bananas, dogs, tortoises, and lizards. All these are under the control, at least partially, of the same social group, and consequently, the two terms being basically synonymous, belong to the same class of beings.[1]

The astrological mythology of the Australians bears the marks of this same mental system. Indeed, this mythology is moulded, as it were, by the totemic organization. Almost everywhere the blacks say that a certain star is a certain particular ancestor.[2] It is more than probable that one might say of such a star, as of the individual with whom it is identified, to which moiety, which marriage class, and which clan it belongs. In so doing, it would be classed in a given group: it would be assigned a kindred and a definite place in society. What is certain is that these mythological

---

[1] We would draw particular attention to this fact, because it offers occasion for a general remark. Whenever a clan or a religious brotherhood exercises magico-religious powers over different kinds of things, it is legitimate to wonder whether this does not indicate a former classification attributing these different kinds of things to this social group.

[2] The sources on this topic are so numerous that we do not cite them all (see Curr 1886–7, vol. I, pp. 255, 403; vol. III, p. 29). This mythology is actually so widespread that Europeans have often believed that the stars were the souls of the dead. [See Louis Rougier, *La Religion astrale des Pythagoriciens*, Paris, 1959, p. 102.—R.N.]

conceptions are found in the Australian societies in which we have found a classification, with all its characteristic features, by moiety and clan; viz. among the Mount Gambier tribes, the Wotjobaluk, and the tribes to the north of Victoria. The sun, says Howitt, is a Krokitch woman of the sun clan who searches every day for her little son who is lost.[1] Bunjil (the star Fomalhaut) was a powerful white cockatoo of the Krokitch moiety before going up into the sky. He had two wives, who, naturally in view of the exogamous rule, belonged to the opposite moiety, Gamutch. They were swans (probably two sub-totems of the pelican). They themselves are also stars.[2]—The Woiworung, cousins of the Wotjobaluk,[3] believe that Bunjil (name of the moiety) went up into the sky in a whirlwind together with his sons[4] who are now all totemic beings (men and animals at the same time); he is Fomalhaut, as among the Wotjobaluk, and each of his sons is a star;[5] two are α and β in the Southern Cross.—Some distance away, the Mycooloon of southern Queensland[6] class the clouds near the Southern Cross under the emu totem; the belt of Orion belongs to the Marbarungal clan, and a falling star to the Jinbabora clan. When one of these stars falls, it strikes a *gidyea* tree and becomes a tree of the same name. This indicates that the tree itself is related to this same clan. The moon is a former warrior, but we are not told his name or the class to which he belonged. The sky is peopled by ancestors from an imaginary epoch.

The same astronomical classifications are employed by the Arunta, whom we shall discuss below in another connexion. For them the sun is a woman of Panunga marriage class,

[1] Howitt 1887, p. 53, fn. 2.

[2] Howitt 1886, p. 415, fn. 1; 1889a, p. 65, fn. 3.

[3] Howitt 1889a, p. 66.

[4] Howitt 1889a, p. 59; cf. p. 63, fn. 2. They correspond to the five fingers.   [5] Howitt 1889a, p. 66.   [6] See Palmer 1884, pp. 293, 294.

and it is the Panunga-Bulthara moiety which is in charge of the religious ceremony concerned with it.[1] It left descendants on earth who are continually reincarnated[2] and form a special clan. But this last detail of the mythical tradition must be a later development. For in the sacred ceremony of the sun the preponderant part is played by individuals belonging to the bandicoot totemic group and to that of the large lizard. This means that the sun must formerly have been a Panunga, of the bandicoot clan, living in the large lizard country. We know, moreover, that this is the case with his sisters. They are merged with him. He is their 'little child', 'their sun'; in short, they are nothing but divisions of him.—The moon, in two different myths, is connected with the opossum clan. In one myth it is a man of this clan;[3] in the other, the moon is itself, but was stolen from a man of the clan,[4] and it was the latter who assigned it its route. We are not told, it is true, to which moiety it belonged. But the clan implies the moiety, or at least implies it in principle among the Arunta.—Concerning the morning star, we know that it belongs to the Kumara class; every evening it goes to hide in a stone in the territory of the 'large lizards', to which it seems to be closely related.[5] In the same way, fire is intimately connected with the euro totem. It was a man of this clan who discovered it in the animal of the same name.[6]

Finally, in many cases where these classifications are not immediately apparent they are nevertheless found, but in a different form from that which we have just described.

[1] Individuals conducting the ceremony must, in the main, be from this moiety (Spencer and Gillen 1899, p. 561).

[2] It is well known that, for the Arunta, each birth is the reincarnation of the spirit of a mythical ancestor (*alcheringa*).

[3] Spencer and Gillen 1899, p. 564.

[4] Spencer and Gillen 1899, p. 565.

[5] Spencer and Gillen 1899, p. 563, bottom.

[6] Spencer and Gillen 1899, p. 446.

Changes have taken place in the social structure which have altered the economy of these systems, but not to the point of making it completely unrecognizable. Moreover, these changes are due in part to the classifications themselves and might thus even reveal them.

What characterizes the latter is that the ideas are organized on a model which is furnished by the society. But once this organization of the collective mind exists, it is capable of reacting against its cause and of contributing to its change. We have seen how species of things, classed in a clan, serve it as secondary or sub-totems; i.e. within the clan a particular group of individuals, under the influence of causes which are unknown to us, comes to feel more specially related to certain things which are attributed, in a general way, to the whole clan. The latter, when it becomes too large, then tends to segment, and this segmentation takes place along the lines laid down by the classification. We must beware of thinking, in fact, that these secessions are necessarily the products of revolutionary or tumultuous movements. More often, indeed, it seems that they have taken place by a completely logical process. It is in this way that, in a large number of cases, the moieties were formed and then split into clans. In many Australian societies they are opposed to each other like the two terms of an antithesis, as black to white,[1] and in the tribes of the Torres Straits as land to sea;[2] moreover, clans formed within each moiety are logically related to each other. Thus in Australia it is rare for the crow to belong to one moiety other than that of thunder, clouds and water.[3] Similarly, when segmentation of a clan becomes necessary, it is individuals grouped around one of the things classed in the clan who detach themselves

[1] See above, p. 22.   [2] See above, p. 28.
[3] This is convincingly shown by a study of the lists of clans divided by moieties given by Howitt (1883, p. 149; 1889a, pp. 52 ff.; 1884b).

from the rest to form an independent clan, and the sub-totem then becomes a totem. Once begun, moreover, the same process may be continued for ever. The sub-clan which emancipates itself in this way takes with it ideologically certain things, other than that used as its totem, which are considered solidary with it. These things play the part of sub-totems in the new clan, and if there is occasion may similarly become centres around which new segmentations may later be produced.

The Wotjobaluk permit us precisely to apprehend this phenomenon in its relations with the classification.[1] Howitt tells us that a certain number of sub-totems are totems in process of formation.[2] 'They gained a sort of independence.'[3] Thus, for certain individuals the white pelican is a totem, and the sun a sub-totem, while others class them in the reverse order. This is probably because these two designations were used for sub-totems of two segments of a former clan, of which the old name was dropped,[4] and which included both the pelican and the sun among the things attributed to it. With time, the two parts detached themselves from their common stem; one took the pelican as principal totem, leaving the sun in second place, while the other did the contrary. In other cases, in which this segmentation cannot be observed so directly, it is expressed in the logical relations which unite sub-clans which have originated from one clan. We shall demonstrate this particularly below, in connexion with certain American societies.[5]

---

[1] It was from this point of view that Howitt studied the Wotjobaluk, and it is this segmentation which, by giving the same kind of things the character sometimes that of totem and sometimes that of a sub-totem, made it difficult to make an exact table of the clans and totems.

[2] Howitt 1889a, p. 63 and particularly p. 64.

[3] Howitt 1885, p. 818.   [4] Howitt 1889a, pp. 63, 64.

[5] See below, pp. 46–7.—This segmentation, and these modifications in the hierarchy of totems and sub-totems resulting from it, may perhaps explain an interesting peculiarity of these social systems. We know that

Now it is easy to see what changes this segmentation must introduce in classification. To the extent that sub-clans which have issued from one original clan preserve the memory of their common origin, they feel that they are relatives, associates, that they are merely parts of the same whole; consequently their totems and the things classed under these totems remain subordinate, in some degree, to the common totem of the whole clan. But with time this sentiment vanishes. The independence of each segment increases, and ends by becoming a complete autonomy. The ties uniting all these clans and sub-clans into the same moiety slacken ever more easily, and the whole society ends up as a scattering of little autonomous groups, all equal among themselves, none subordinate to another. Naturally, the classification is changed in consequence. The kinds of things attributed to each of these sub-divisions constitute as many separate genera, all on the same level. All sign of hierarchy has disappeared. It may easily be conceived, however, that traces of it should still remain within each of these small clans. Beings connected with the sub-totem, which has now become a totem, continue to be subsumed under the latter. But, in the first place, they can no longer be so many, given the fissive character of these little groups. Furthermore, however little this character is realized, each sub-totem will end up by being elevated to the dignity of a totem, and every species and every subordinate variety will have become a major genus. So the old classification will have given place to a simple division without any internal organization, a division of things *per capita* and no longer by origins. But, at the same time, as it is made be-

totems, particularly in Australia, are very commonly animals, and that they are much more rarely inanimate objects. It may be that originally they were all taken from the animal world. But inanimate objects were classed under these primitive totems, which, following the segmentation, ended up by being promoted to the rank of principal totems.

tween a considerable number of groups, it will be found to cover practically the entire universe.

Arunta society is in this position. They have no complete classification, no integrated system. But, as we read in the very words of Spencer and Gillen, 'in fact, there is scarcely an object, animate or inanimate, to be found in the country occupied by the natives which does not give its name to some totemic group of individuals'.[1] We find fifty-four species of things mentioned in their work as totems of as many totemic groups; and furthermore, as the authors themselves were not concerned to draw up a complete list of these totems, that which we have compiled from scattered indications in their book is certainly not exhaustive.[2] Now the Arunta tribe is surely one of these in which the process of segmentation has been carried to its very limit; for, following changes which have occurred in the structure of this society, all obstacles capable of keeping it in check have disappeared. Under the influence of causes which have been described elsewhere,[3] the totemic groups of the Arunta

---

[1] Spencer and Gillen 1899, p. 112.

[2] It may be helpful if we reproduce this list. Naturally, we follow no order in our enumeration. Wind, sun, water or cloud (p. 112), rat, *witchetty* grub, kangaroo, lizard, emu, *hakea* flower (p. 116), eagle-hawk, *elonka* (a fruit), a kind of manna, wildcat, *irriakura* (kind of bulb), the grub of a butterfly, bandicoot, *ilpirla* manna, honey-ant, frog, *chankuna* berry, plum tree, *irpunga* fish, opossum, wild dog, euro (pp. 167 ff.), little night hawk (p. 232), carpet snake (p. 242), large white bat (p. 299), little grub (p. 302), grass seed (p. 311), *interpitna* fish (p. 316), *coma* snake (p. 317), the native pheasant, a kind of Marsdenia fruit (p. 320), jerboa (p. 329), evening star (p. 360), large lizard, small lizard (p. 389), small rat (pp. 389, 395), *alchantwa* seed (p. 390), another kind of small rat (p. 396), small hawk (p. 397), *okranina* snake (p. 399), wild turkey, magpie, white bat, little bat (p. 405). There are also the clans of a certain kind of seed and the large beetle (p. 411), *inturrita* pigeons (p. 410), water-beetle (p. 414), hawk (p. 416), quail, bull-dog ant (p. 417), two sorts of lizards (p. 439), nail-tailed wallaby (p. 441), another kind of *hakea* flower (p. 444), the fly (p. 546), and the bell-bird (p. 635).

[3] *Année Sociologique*, vol. V, 1902, pp. 108 ff.

were led very early to leave the natural framework in which they were formerly confined, and which served them in a way as a skeleton; viz. the framework of the moiety. Instead of remaining strictly located in a particular half of the tribe, each of them spread freely throughout the whole extent of the society. Having thus become foreign to the regular social organization, and fallen almost to the level of private associations, they were able to multiply and to crumble almost *ad infinitum*.

This crumbling still continues. There are indeed species of things whose rank in the totemic hierarchy is still uncertain, as Spencer and Gillen assert; it is not known whether they are principal totems or sub-totems.[1] This means that the groups are still in a mobile state, like the Wotjobaluk clans. On the other hand, there sometimes exist links between totems at present assigned to independent clans which show that formerly they must have been classed in the same clan. Such is the case with the *hakea* flower and the wild cat. Thus the marks engraved on the *churinga* of wild-cat men represent, and only represent, trees with *hakea* flowers.[2] According to myths, wild cats used to feed on the *hakea* flower in olden times; and the original totemic groups are generally reputed to have lived on their totems.[3] This means that these two sorts of things have not always been strangers to each other, but became so only when the single clan which comprised them segmented. The plum-tree clan also seems to derive from this same

[1] Thus Spencer and Gillen are not quite sure whether the rock pigeon is a totem or a secondary totem (pp. 410 and 448). Similarly, the totemic value of various species of lizards is not determined: thus the mythical beings who created the first men to belong to the lizard totem then changed themselves into another kind of lizard (p. 389).

[2] Spencer and Gillen 1899, pp. 147–8. [In fact, certain of the designs also represent the tracks of men dancing round the *hakea* trees, and the sticks beaten to keep time in the dancing.—R. N.]

[3] P. 449.

complex clan: flower-wild cat.[1] Different species of animals
and other totems, notably that of the small rat,[2] detached
themselves from the lizard totem.[3] We may therefore be
sure that the primitive organization underwent an extensive process of dissociation and segmentation which has not
yet ended.

If, then, we no longer find a complete system of classification among the Arunta, this is not because there has never
been one: it is because it disintegrated in company with the
fragmentation of the clans. The state in which it is found
only reflects the present state of the totemic organization in
this same tribe; it is a further proof of the close relation
which unites these two orders of facts. Moreover, it has not
disappeared without leaving visible signs of its former
existence. We have already noted survivals in Arunta
mythology. But a better demonstration still is the way in
which things are distributed among the clans. Very often,
other kinds of things are connected with the totem, just as
in the complete classifications that we have examined. This
is a last vestige of subsumption. Thus the frog clan is
specially associated with the gum tree;[4] the water hen is
connected with water.[5] We have already seen that there is a
close relation between the water totem and fire: on the
other hand, with fire are connected eucalyptus branches, the
red flowers of the Eremophila,[6] the sound of a horn, heat,
and love.[7] The beard is associated with the jerboa rat totems,[8]
and eye diseases with the fly totem.[9] The most frequent case
is that in which the creature thus related to the totem is a

[1] Pp. 283, 299, 403, 404. [2] P. 441.

[3] Pp. 150, 440.

[4] The *churinga*, individual emblems in which the souls of the ancestors
are thought to reside, bear representations, in the frog clan, of gum
trees; ceremonies in which the clan myths are acted out include the
drawing of a tree and its roots (pp. 145, 147, 625, 626; cf. pp. 325, 344
and Figs. 72 and 74). [5] P. 448.

[6] Pp. 238, 321. [7] P. 545. [8] P. 329. [9] P. 546.

bird.[1] A little black bird, Alatipa, has as mates the honey ants which live, as it does, on *mulga* bushes,[2] and so does another little bird, Alpirtaka, which seeks out the same inhabitants.[3] A species of bird called Thippa-Thippa is allied to the lizard.[4] The plant called Irriakura has the ring-necked parrot as its partner.[5] People of the *witchetty* grub do not eat certain birds which are called their companions (*quathari*, which Spencer and Gillen translate as 'mates').[6] The kangaroo totem has two kinds of birds subordinate to it,[7] and the same is the case with the euro.[8] What clinches the demonstration that these connexions are indeed the remains of a former classification is that the creatures which are thus associated with others were once of the same totem. The Kartwungawunga birds were formerly, according to legend, Kangaroo men and used to eat kangaroo. The two species of bird connected with the honey-ant totem were formerly honey-ants. The Unchurunqa, beautiful little red birds, originally belonged to the euro clan. The four species of lizards are composed of two pairs, in each of which one is simultaneously the associate and the transformation of the other.[9]

Lastly, a final proof that we are indeed dealing with an altered form of earlier classifications among the Arunta is that the series of intermediate states may be discovered, almost without break in continuity, by which this organization is connected with the classic Mount Gambier type. Among the Chingalee,[10] northern neighbours of the Arunta

[1] Spencer and Gillen speak only of birds. But in fact the situation is much more general.   [2] Pp. 448, 447.

[3] Pp. 448, 188, 646. Note the similarity between their names and that of Ilatirpa, the great ancestor of this totem.

[4] P. 305. In certain clan ceremonies two individuals representing two birds of this species dance round the 'lizard'. And, according to the myths, this dance was already performed at the time of the Alcheringa.

[5] P. 320. Cf. pp. 318, 319.   [6] Pp. 447, 448.   [7] P. 448.

[8] P. 448.   [9] Pp. 448, 449.   [10] See Mathews 1900.

inhabiting the northern territory of Australia (Gulf of Carpentaria), we find, just as among the Arunta themselves, an extreme dispersion of things among very numerous— i.e. very fragmented—clans; fifty-nine different totems are reported. Just as among the Arunta, the totemic groups have ceased to be classed under moieties; each of them overlaps the two moieties into which the tribe is divided. But the diffusion in this case is not so complete. Instead of being distributed at random and irregularly throughout the society, they are allocated to particular groups according to fixed and localized principles, even though the groups belong to different moieties. Each moiety is divided into four marriage classes;[1] and each class of one moiety may marry only a particular class of the other, which has or may have the same totems as the first. Together, these two corresponding classes thus contain a definite group of totems and things which are not found elsewhere. For example, there belong to the two classes Choongoora-Chabalye pigeons of all kinds, ants, wasps, mosquitoes, centipedes, bees, grass, grasshopper, various snakes, and so on; certain stars, the sun, clouds, rain, water-hen, ibis, thunder, eagle-hawk, brown

[1] [Durkheim and Mauss write that each moiety is divided into eight marriage classes, but Mathews says clearly that each 'is subdivided into four sections, making a total of eight divisions in the community' (p. 494).—R. N.] Further on this point, there is a remarkable kinship between this tribe and the Arunta, who also have eight marriage classes; at least this is so among the northern Arunta, and among the others the same sub-division of the original four classes is in process of formation. The cause of this segmentation is the same in both societies, viz. a change from matrilineal to patrilineal descent. It has already been shown how this change would make any marriage impossible if the initial four classes did not sub-divide (*Année Sociologique*, vol. V, 1902, p. 106, fn. 1).—Among the Chingalee, this change has been effected in a very special way. The moiety, and consequently the marriage class, continues to be matrilineal; only the totem is inherited from the father. This explains how it is that each class of one moiety has in the other a corresponding class with the same totems. It is because a child belongs to a class of the mother's moiety; but it has the same totems as its father, who belongs to a class of the other moiety.

hawk, black duck, etc. are attributed to the group formed of the classes Chowan and Chowarding; the wind, lightning, the moon, frogs, etc. to the Chambeen-Changalla group; shell-fish, the *bilbi* rat, crow, porcupine, kangaroo, etc. to the Chagarra-Chooaroo group. Thus things are still, in one sense, ordered by fixed categories, but these latter are already a little more artificial and less consistent, since each of them is formed of two sections which belong to two different moieties.

With another tribe of the same area we go one step further towards organization and systematization. Among the Moorawaria of the Culgoa river,[1] the segmentation of the clans is pushed still further than among the Arunta; we are told, in fact, of 152 species of objects which serve as totems to as many different clans. But this innumerable multitude of things is ordered in a regular fashion by the two moieties, Ippai-Kumbo and Kubi-Murri.[2] We are thus very close, save for the extreme fragmentation of the clans, to the classical type. It is necessary only that the society, instead of being dispersed to this extent, should be concentrated; that the clans, which are so separated, should reunite according to their natural affinities in such a way as to form larger groups; that, consequently, the number of principal totems should diminish (other things, which at present serve as totems, assuming a subordinate place to the preceding)—and we arrive exactly at the Mount Gambier systems.

To sum up, though we may not be justified in saying that this way of classifying things is necessarily implied by totemism, it is certain, in any case, that it is found very often in societies organized on a totemic basis. There is thus

[1] Mathews 1898.

[2] No special names are known designating the moieties in this tribe. We therefore call each of them by the names of its two marriage classes. It will be noted that the nomenclature is that of the Kamilaroi system.

a close link, and not an accidental relation, between the social system and this logical system. We shall now see how other forms of classification, presenting a higher degree of complexity, may be connected to this primitive form.

Chapter Three

ZUÑI, SIOUX

ONE OF THE MOST remarkable examples is offered by the Zuñi.[1]

Powell writes that the Zuñi 'represent an unusual development of the primitive concepts concerning the relations of things'.[2] Among them, the idea which the society has of itself, and its world-view, are so interlaced and merged that their organization has perfectly exactly been described as 'mytho-sociologic'.[3] Cushing does not exaggerate, therefore, when he says of his studies among this people that: 'I have

[1] The Zuñi have been admirably studied by Cushing (1883; 1896). He says they are at once 'the most archaic' and 'the most highly developed' of the Pueblo peoples (1896, p. 325). They make admirable pottery, grow wheat and peaches imported by the Spaniards, and are famous jewellers; for nearly two hundred years they have been in contact with the Mexicans. Today they are Catholics, but only outwardly; they have retained their rites, customs and beliefs (p. 335). They live all together in a pueblo, i.e. a single town, formed in reality of six or seven houses rather than six or seven groups of houses. They are thus characterized by an extreme social concentration and a remarkable conservatism, as well as by a great faculty for adaptation and evolution. Though they are not primitives as described by Cushing and Powell (1896, p. lvii; 1883, p. xxvii) it is certain that we are dealing with a type of thought which has developed in accordance with very primitive principles.

The history of this tribe is summed up by Cushing (1896, pp. 327 ff.); but the hypothesis which he proposes, according to which the Zuñi are of a dual origin, does not seem to us at all proved.

[2] Powell 1896, p. lix.   [3] Cushing 1896, pp. 367, *passim*.

become convinced that they bear on human history . . . for the Zuñis, say, with all their strange, apparently local customs and institutions and the lore thereof, are representative in a more than merely general way of a phase of culture. . . .' And he congratulates himself on the fact that contact with them should have widened his understanding of 'the earliest conditions of man everywhere as nothing else could have done'.[1]

Indeed, what we find among the Zuñi is a veritable arrangement of the universe.[2] All beings and facts in nature, 'the sun, moon, and stars, the sky, earth and sea, in all their phenomena and elements; and all inanimate objects, as well as plants, animals, and men', are classed, labelled, and assigned to fixed places in a unique and integrated 'system' in which all the parts are co-ordinated and subordinated one to another by 'degrees of resemblance'.[3]

In the form in which we now find it, the principle of this system is a division of space into seven regions: north, south, west, east, zenith, nadir, and the centre. Everything in the universe is assigned to one or other of these seven regions. To mention only the seasons and the elements, the wind, breeze or air, and the winter season are attributed to the north; water, the spring and its damp breezes, to the west; fire and the summer, to the south; the earth, seeds, the frosts which bring the seeds to maturity and end the year, to the east.[4] The pelican, crane, grouse, sagecock, the evergreen oak, etc. are things of the north; the bear, coyote, and spring grass are things of the west. With the east are classed

---

[1] Cushing 1896, p. 378.   [2] Cushing 1896, p. 370.

[3] Cushing 1883, p. 9. According to Cushing, 'the degrees of relationship seem to be determined largely, if not wholly, by the degrees of resemblance'. Elsewhere (1896, pp. 368, 370) the author believes it possible to employ his system of explanation with complete rigour; but, as far as the Zuñi are concerned, we must be more careful. We shall in fact demonstrate the arbitrariness of these classifications.

[4] Cushing 1896, pp. 369–70. Seeds were formerly placed to the south.

the deer, antelope, turkey, etc. Not only things, but social functions also are distributed in this way. The north is the region of force and destruction; war and destruction belong to it; to the west, peace (as we render the word 'war cure', which we do not quite understand), and hunting; to the south, the region of heat, agriculture and medicine; to the east, the region of the sun, magic and religion; to the upper world and the lower world are assigned diverse combinations of these functions.[1]

A particular colour is attributed to each region and characterizes it. The north is yellow because, it is said,[2] the light is yellow when the sun rises and sets; the west is blue because of the blue light that is seen at sunset.[3] The south is red because it is the region of summer and fire, which is red. The east is white because it is the colour of the day. The upper regions are streaked with colours like the play of light among the clouds; the lower regions are black like the depths of the earth. As for the centre, the navel of the world, representative of all the regions, it is all the colours simultaneously.

So far, it seems that we are in the presence of a classification which is quite different from those which we have first examined. But there is something which allows us to suppose that there is a close link between the two systems, viz. that *this division of the world is exactly the same as that of the clans within the pueblo*. This also 'is divided, not always very clearly to the eye, but very clearly in the estimation of the people themselves, into seven parts, corresponding, not perhaps in arrangement topographically, but in sequence, to

[1] Cushing 1896, pp. 371, 387, 388.

[2] We report these explanations without committing ourselves to their validity. The reasons behind the distribution of colours are probably even more complex. But the reasons given are not without interest.

[3] Cushing says that it is because of the 'blue of the Pacific', but he does not establish that the Zuñi have ever seen the ocean.

their subdivisions of the "worlds". . . Thus, one division of the town is supposed to be related to the north. . . .; another division represents the west, another the south,' etc.[1] The relationship is so close that each of the quarters of the pueblo has its characteristic colour, as do the regions; and this colour is that of the corresponding region.

Now each of these divisions is a group of three clans, except that which is situated at the centre and has only one, and 'These clans are also, as are those of all other Indians, totemic.' [2] We give here the complete table of them, for there will be occasion to refer to them in order to understand the observations which follow:[3]

region	*clans*
north	crane, or pelican
	grouse, or sagecock
	yellow wood, or evergreen oak (clan almost extinct)
west	bear
	coyote
	spring-herb
south	tobacco
	maize
	badger
east	deer
	antelope
	turkey
zenith	sun (extinct)
	eagle
	sky
nadir	frog, or toad
	rattlesnake
	water
centre	macaw, the clan of the perfect centre

[1] Cushing 1896, p. 367.

[2] Cushing 1896, p. 368. Descent is matrilineal; the husband resides at his wife's place.   [3] Cushing 1896, p. 368.

The relation between the distribution of the clans and that of things according to region appears even more clearly when it is recalled that, in a general fashion, whenever we find different clans grouped together in such a way as to form a whole with a certain moral unity, we may be practically certain that they have segmented from the same original clan. If then this rule is applied to the Zuñi, we see that there must have been a time in the history of this people when each of these six groups of three clans constituted a single clan, from which it follows that the tribe was divided into seven clans,[1] corresponding exactly to the seven regions. This hypothesis, which is already very probable for the general reason given, is moreover expressly confirmed by an oral document whose antiquity is certainly considerable.[2] There is given in it a list of the six great priests who represent the six groups of clans in the important religious brotherhood called after the 'knife'. Now the priest who is master of the north is called the *first in the kin of the bear*; that of the west, the *first in the kin of the coyote*; that of the south, *first in the kin of the badger*; that of the east, *first in the kin of the turkey*; that of above, *first in the kin of the eagle*; that of the below, *first in the kin of the snake*.[3] If we refer to the list of clans, we see that the six animals to whose kindreds the six great priests belong serve as totems to six clans, and that these six clans are oriented exactly as are the corresponding creatures, with the sole exception of

[1] By counting the clan of the centre and regarding it as forming a separate group, apart from the two moieties of three clans—which is doubtful.

[2] The text is in verse; and versified texts are preserved far better than texts in prose. It is certain, moreover, that at the time of their conversion, i.e. in the eighteenth century, the Zuñi had an organization very similar to that which Cushing studied among them. Most of the brotherhoods and clans existed in an absolutely identical form, as may be established on the basis of the names inscribed in the baptismal registers of the mission (Cushing 1896, p. 383).

[3] Cushing 1896, p. 418.

the bear, which in the more recent classifications is classed among things of the west.[1] They thus belong (with this single exception) to as many different groups. Consequently, each of these clans is invested with a veritable primacy within its own group; it is evidently considered as its representative and chief, since the person charged with this representation is taken from it. This is to say that it is the primary clan from which the other clans of the group are derived by segmentation. It is a general fact among the Pueblo Indians (and also elsewhere) that the first clan in a phratry is also its original clan.[2]

More than this, not only do the division of things by regions and the division of society by clan correspond, but they are inextricably interwoven and merged. One may equally well say that things are classified either with the north, south, etc., or with the clans of the north, of the south, etc. This is particularly evident in the case of totemic animals; they are manifestly classed in their clans, as well as with particular regions.[3] The same is the case with all things, and even with social functions. We have seen how they are distributed between the regions;[4] and this distribution reduces itself in reality to a division between the clans. These functions, in fact, are at present exercised by religious brotherhoods which, in everything concerning these different offices, have been substituted for the clans. Now the brotherhoods are recruited, if not solely at least principally,

[1] It is probable that this clan changed orientation over time.

[2] As we are concerned for the moment only with the demonstration that the six groups of three clans were produced by the segmentation of six original clans, we shall leave aside the question of the nineteenth clan. We shall return to this matter below [p. 53, n. 3].

[3] Thus the priest-fathers decided that the creatures and things of summer and the southern space belonged to the People of Summer: '. . . those of winter and the northern space, to the People of Winter', etc.

[4] We use this term in abbreviation for regions of orientation.

from the clans attributed to the same regions as the corresponding functions.[1] Thus the societies of the knife, glass wand, and cactus, which are brotherhoods of war, are grouped 'not in an absolutely rigorous manner, but in principle' with the clans of the north; the people of priesthood, the bow, and the hunt are taken from the clans of the west; 'priests of the priesthood' are taken from those of the east, as well as those of the cotton-bloom and the monster-bird which form the society of the great dramatic dance (magic and religion); and the societies of the great fire or embers, whose functions are not explicitly reported but which certainly have to do with agriculture and medicine,[2] are taken from the clans of the south. To be exact, we cannot say that things are classified by clans, or by quarters, but by oriented clans.

This system, then, is far from being separated by an abyss from the Australian. However different in principle a classification by clans may be from one by quarters, among the Zuñi they are superimposed one on the other and agree exactly. We can go even further. A number of facts show that it is the classification by clans which is the older, and that this was the model on which the other was formed.

Firstly, the division of the world by quarters has not always been what it is. It has a history, the principal phases of which can be reconstructed. Before the division into seven, there was certainly one into six, and its traces may still be found.[3] And before the division into six, there was one into four, corresponding to the four cardinal points. This

[1] Cushing 1896, pp. 371, 387–8.

[2] Everywhere in America there is a relation between heat, particularly that of the sun, and agriculture and medicine.—As for the brotherhoods included in the regions of above and below, their functions are generation and the preservation of life.

[3] We know that the notion of 'centre' is of relatively late development. The centre 'was found at a particular time' (Cushing 1896, pp. 388, 390, 398, 399, 403, 424–30).

is doubtless the explanation of the fact that the Zuñi should have distinguished only four elements, situated in four regions.[1]

Now it is at very least remarkable that with these variations in the classification by quarters there are other corresponding variations, exactly parallel, in the classification by clans. There is often question of a division into six clans which was evidently anterior to the division into seven: this is why the clans from which are chosen the great priests who represent the tribe in the brotherhood of the knife number six. Finally, the division into six was itself preceded by a division into two primary clans or moieties which comprised exhaustively the totality of the tribe: this fact will be established below.[2] Now the division of a tribe into two moieties corresponds to a table of the quarters divided into four parts. One moiety occupies the north, another the south, and between, in order to separate them, there is the line running from east to west. We shall see distinctly, among the Sioux, the relation uniting this social organization to this distinction of the four cardinal points.

Secondly, one fact which shows well that the classification by quarters was superimposed more or less late upon the classification by clans is that it has been adapted to the latter only clumsily and with the aid of a compromise. If the principle on which the former system is based were adhered to, each kind of thing ought to be classified completely in one and only one particular region; for example, all eagles should belong to the upper region. Now the Zuñi knew that there were eagles in every region. It was proposed, then, that each species had a predilection for a particular region;

---

[1] Cushing 1896, p. 369. The following passages demonstrate the point well: 'They carried the tubes of hidden things . . . like the regions of men, four in number. And the revealing-balls thereof, . . . like the regions of men, four in number.'

[2] See below, pp. 53–4.

that there, and there only, it existed in its highest and perfect form. But at the same time it was supposed that this same species had representatives, only smaller and less excellent, in the other regions, and that these were distinguished from each other in that each was of the colour characteristic of the region to which it was attributed: thus, besides the eagle placed at the zenith, there are fetish eagles for all the regions; the yellow eagle, blue eagle, white eagle, and black eagle.[1] Each of them has in its own region all the virtues attributed to the eagle in general. It is not impossible to reconstruct the course by which Zuñi thought arrived at this complex conception. In the beginning things were divided by clans; each animal species was then assigned entirely to a certain clan. This total attribution occasioned no difficulty, for there was no contradiction in conceiving a whole species as standing in a relation of kinship to one or other human group. But when the classification by quarters was established, a downright impossibility appeared; the facts were too clearly opposed to a rigorously exclusive localization. It was thus absolutely necessary that the species, though remaining pre-eminently concentrated at a unique point, as in the former system, should however be diversified in order to be able to be dispersed, under secondary forms and various aspects, in all directions.

Thirdly, it is reported that in many cases things are, or were at a certain time in the past, directly classified under the clans and are related only indirectly through these to their respective quarters.

First of all, so long as the six initial clans were still undivided, the things which have since become the totems of the new clans which have formed obviously had to belong to the initial clan as sub-totems, subordinate to the totem of that clan. They were species of it.

[1] Cushing 1883, pp. 18, 24, 25, Pls. III–VI.

The same immediate subordination is still found today in the case of a particular category of creatures, viz. game. All species of game are divided into six classes, and each of these classes is considered as placed in dependence on a particular 'prey animal'. Each of the animals to which this prerogative is attributed inhabits one region. They are: to the north, the mountain lion, which is yellow; west, the bear, which is dark; south, the badger, which is black and white;[1] east, the white wolf; at the zenith, the eagle; at the nadir, the prey mole, black as the depths of the earth. Their souls live in little collections of stones which are believed to be their forms, and which on occasion are painted with their characteristic colours.[2] For example, the coyote, the mountain sheep, etc., are subject to the bear.[3] If it is desired to assure plentiful coyote-hunting or to maintain the specific power of the species, it is the bear fetish which is used in certain special rites.[4] Now it is noteworthy that of these six animals three are still used as totems by existing clans and are oriented as these clans themselves are; these are the bear,

[1] The reasoning by which the Zuñi justify this attribution of the badger shows how much these associations of ideas depend on causes which are quite foreign to the intrinsic nature of the associated things. The south has the colour red and it is said that the badger belongs to the south because, on the one hand, it is black and white, and on the other, red is neither white nor black (Cushing 1883, p. 17). Here we see ideas which are related in a logic singularly different from our own. [The text actually says: 'for thy coat is ruddy and marked with black and white equally, the colours of the land of summer, which is red, and stands between the day and the night, and thy homes are on the sunny sides of the hills'.—R. N.]   [2] Cushing 1883, p. 15.

[3] The distribution of game among the six prey animals is set out in a number of myths (Cushing 1883, p. 16) which do not agree in all their details but are based on the same principles. These disagreements are easily explained by modifications which have taken place in the orientation of the clans.

[4] The six fetish animals coincide exactly, except for two, with the six 'prey animals' of the myths. The divergence is due simply to the fact that two species were replaced by two others which were related to the former.

the badger, and the eagle. Also, the mountain lion is only a substitute for the coyote, which was formerly the totem of one of the northern clans.[1] When the coyote moved over to the west, it left as its replacement in the north one of the species related to it. There was thus a time when four of these privileged animals were totemic. As for the prey mole and the white wolf, it should be observed that none of the creatures serving as totems to the clans of the two corresponding regions (east and the nadir) is a prey animal.[2] It was therefore necessary to find substitutes for them.

Thus the different sorts of game are conceived as directly subordinate to the totems or to their substitutes. It is only through the latter that they are connected with their respective quarters. Which is to say that the classification of things by totems, i.e. by clans, preceded the other.

There is still another point of view from which the same myths indicate this anteriority of origin. The six prey animals are not only set over game, but over the six regions: one of the six parts of the world is assigned to each of them and is guarded by it.[3] It is through their mediation that creatures placed in their regions communicate with the god who created mankind. The region, and everything belonging to it, is thus seen as being in a certain relation of dependence upon the animal totems. This could never have come about if the classification by quarters had been the earlier.

Thus, beneath the classification by regions, which alone was apparent at first, we find another which is identical at all points with that which we have seen already in Australia.

[1] This is proved by the fact that the fetish of the yellow coyote, which is assigned as secondary species to the north, nevertheless takes precedence over the blue coyote fetish, which belongs to the west (Cushing 1883, pp. 26, 31).

[2] There is indeed the snake which is the totem of the nadir, and which, according to present ideas, is a prey animal. But it is not so for the Zuñi. For them, prey animals can only be those with claws.

[3] Cushing 1883, pp. 18, 19.

This identity is even more complete than appears from the foregoing. Not only were things directly classified at one time by clans, but these clans themselves were classed in two moieties as in Australian societies. This emerges from the evidence in a myth recorded by Cushing.[1] The first great priest and magician, say the Zuñi, brought two pairs of eggs to mankind just after they had been created; one was a marvellous dark blue like the sky; the other was dark red, like mother earth. He said that one was summer and the other winter, and he invited men to choose. The first to make their choice decided on the blue ones; they were delighted that the young birds had no feathers. But when these grew up they became black: they were ravens, whose descendants, veritable scourges, left for the north. Those who chose the red eggs saw the birth of the brilliant macaw parrot; they shared seeds, warmth, and peace. 'Thus,' the myth continues, 'first was our nation divided into the People of Winter and the People of Summer. . . .' Some became 'the macaw and the kindred of the macaw, the Múla-kwe; whilst those who had chosen the ravens became the Raven-people, or Ka'kâ-kwe'.[2] Thus the society began by being divided into two moieties, one situated to the north, the other to the south; one had as its totem the raven, which has disappeared, the other the macaw, which still exists.[3] Mythology still preserves the memory of the sub-division of each moiety into clans.[4] According to their respective natures, tastes, and aptitudes, the people of the

[1] Cushing 1896, pp. 384 ff.

[2] The word Ka'kâ-kwe seems indeed to be the old name for the raven. If this identification is admitted, it settles all the problems raised by the etymology of the word and the origin of the feast of the Ka'kâ-kwe. See Fewkes 1897, p. 265, fn. 2.

[3] The macaw clan, which is now the only one belonging to the region of the centre, was thus originally the first clan, the clan of origin of the summer moiety. [But cf. p. 48, n. 3.—R. N.]

[4] Cushing 1896, p. 386; cf. pp. 405, 425–6.

north or the raven became, says the myth, people of the bear, people of the coyote, deer, crane, etc., and similarly with the people of the south and the macaw. And once they were established, the clans shared the essences of things: for example, the seeds of hail and snow belonged to the elk; the seeds of water, etc. belonged to the toad clans. Here we have a new proof that in the beginning things were classed by clans and by totems.

The foregoing permits the conclusion that the Zuñi system[1] is really a development and a complication of the

[1] We speak of the Zuñi system because it has been the best and the most completely observed among them. We cannot establish in an absolutely certain fashion that the other Pueblo Indian systems were the same; but we are convinced that studies being carried out at present by Fewkes, Bourke, Stevenson, and Dorsey will lead to similar results. What is certain is that among the Hopi of Walpi and Tusayan there are nine groups of clans, similar to those that we have seen among the Zuñi; the first clan in each of these groups has the same name as the whole group, proving that this group is the result of segmentation of the initial clan (Mindeleff 1891, p. 12). These nine groups include an innumerable multitude of sub-totems which seem indeed to cover everything in nature. Also, there is explicit mention of clans with mythically fixed orientations. Thus the rattlesnake clan came from the west and the north, and it comprises a certain number of things which consequently are also oriented: different sorts of cactus, pigeons, marmots, etc. From the east came the group of clans with the horn as totem, including the antelope, deer, and mountain sheep. Each group originated in a clearly oriented region. Furthermore, the colour symbolism corresponds well with that which we have observed among the Zuñi (Fewkes 1897, pp. 276 ff.; cf. Mallery 1886, p. 56). Finally, just as among the Zuñi, prey monsters and game are distributed among the regions .There is a difference, though, in that the regions do not correspond to the cardinal points.
The ruined pueblo of Sia seems to have preserved a very clear memory of this state of collective thought (Stevenson 1894, pp. 28, 29, 32, 38, 41). That things there were divided first of all by clans, and then by regions, is well shown by the fact that in each region there is a representative of each divine animal. But at the present day the clans no longer exist except as survivals.
We believe that similar methods of classification will be found among the Navaho (Matthews 1887, pp. 448–9; cf. Buckland 1893, p. 349). We also think, though we cannot prove it here, that many facts in the

Australian system. But what finally demonstrates the reality of this relationship is that it is possible to discover the intermediate stages connecting these extremes, and thus to discern how the one developed from the other.

The Omaha tribe of the Sioux, described by Dorsey,[1] are precisely in this mixed position: the classification of things by clans is still very clear, and was formerly even clearer, but the systematic idea of regions is only in process of formation.

The tribe is divided into two moieties, each containing five clans. These clans are recruited by exclusively patrilineal descent; which is to say that totemic organization, properly speaking, and the cult of the totem are in decline.[2] Each of these is sub-divided in its turn into sub-clans which themselves are sometimes further sub-divided. Dorsey does not say that everything in the world is divided among these different groups. But if the classification is not exhaustive, and perhaps never really was, certainly it must have been very comprehensive, at least in the past. This is shown by a study of the only complete clan which has been preserved for us;[3] this is the Chatada clan, which is part of the first moiety. We shall leave on one side other accounts which are probably mutilated, and which in any case give us the

[1] Dorsey 1884, pp. 211 ff.; 1894; 1896. Cf. the Teton, Omaha, and Osage texts published in *Contributions to North American Ethnology*, vol. III, part 2, and vol. VI, part 1; Kohler 1897.

[2] Generally speaking, where descent is patrilineal the totemic cult weakens and tends to disappear (Durkheim 1898, p. 23). Dorsey actually mentions the decadence of the totemic cults (1894, p. 371).

[3] Dorsey 1896, p. 226. It seems fairly likely to us that this clan was the bear clan; this is in fact the name of the first sub-clan. Moreover, the clan corresponding to it in the other Sioux tribes is a bear clan.

symbolism of the Huichol (see the review of Lumholtz 1900, in *Année Sociologique*, vol. VI, 1903, pp. 247–53) and that of the Aztecs, 'those other Pueblos' as Morgan writes (1877, p. 199), might be decisively explained by facts of this kind. This idea, moreover, has been expressed by Powell, Mallery, and Cyrus Thomas.

same phenomena but with a lesser degree of complication.

The meaning of the word used to designate this clan is uncertain; but we have a fairly full list of the things which are connected with it. It comprises four sub-clans, which are themselves segmented.[1]

The first sub-clan is that of the black bear. It comprises the black bear, the raccoon, the grizzly bear, and the porcupine, which seem to be totems of the segments.

The second is 'they who do not eat (small) birds'. Under it come: (1) hawks; (2) blackbirds, which are themselves divided into those with white heads, red heads, and yellow heads, and those with red wings; (3) grey blackbirds, or 'Thunder people', who in turn are sub-divided into meadow larks and prairie chickens; and (4) owls, themselves divided into large, medium and small.*

The third sub-clan is that of the eagle; it comprises in the first place three kinds of eagle; and a fourth segment which is called 'Workers' and appears not to be related to a particular order of things.

Lastly, the fourth sub-clan is that of the turtle. It is related to the fog, which its members have the power to stop.[2] Four particular species of the same animal are subsumed under the genus turtle.

Since we may justifiably believe that this case was not unique, and that many other clans must have possessed similar divisions and sub-divisions, it is not a bold supposi-

---

[1] Dorsey 1884, pp. 236 ff.—Dorsey uses the words 'gens' and 'sub-gens' to designate these groupings. It does not seem necessary to us to adopt a new term to designate clans with patrilineal descent. They are only a species of the genus.

* [Actually, 'Owl and Magpie People', comprising great owls, small owls, and magpies.—R. N.]

[2] The fog is certainly represented in the form of a tortoise. We know that among the Iroquois, fog and storm belong to the hare clan (cf. Frazer 1899, p. 847).

tion that the system of classification still to be observed among the Omaha was once more complex than it is today. Now besides this distribution of things, analogous to that reported from Australia, we can see the apparition, though in a rudimentary form, of notions of orientation.

When the tribe camps, the encampment is made in a circular form; and within this circle each particular group has a fixed place. The two moieties are respectively to the right and the left of the route followed by the tribe, the ascription of sides being made with reference to the point of departure. Within the semicircle occupied by each moiety, the clans, in their turn, are clearly localized with respect to each other, and the same is the case with the sub-clans. The places thus assigned to them depend less on their relationships to each other than on their social functions, and consequently on the nature of the things subordinate to them and over which their influence is thought to be exercised. Thus in each moiety there is a clan which stands in a special relationship to thunder and war; one is the elk clan, the other that of the Ictasandas. They are placed facing each other at the camp entrance, more ritual than real, which they guard;[1] and it is by relation to them that the other clans are disposed, still according to the same principle. Things are thus distributed in this way within the camp at the same time as the social groups to which they are attributed. Space is shared among the clans, and among beings, events, etc. which belong to these clans. But it is clear that what is divided in this way is not cosmic space, but only the space occupied by the tribe. Clans and things are orientated, not yet according to the cardinal points, but with reference to the centre of the camp. The divisions do

[1] Fletcher 1898, p. 438.—This disposition is adopted only during general movements of the tribe (Dorsey 1884, pp. 219 ff., 286, §133; cf. 1896, p. 226).

not correspond to the quarters properly speaking, but to ahead and behind, right and left, with respect to this central point.[1] Moreover, these particular divisions are attributed to the clans, rather than the clans being attributed to them as is the case among the Zuñi.

In other Sioux tribes the idea of orientation becomes more distinct. Like the Omaha, the Osage Indians are divided into two moieties, one situated to the right, the other to the left;[2] but whereas among the former the functions of the two moieties merged at certain points (we have seen that each had a clan of war and the thunder), here they are clearly distinguished. One half of the tribe is in charge of war, the other of peace. This necessarily results in a more exact localization of things. We find the same organization among the Kansa Indians. Moreover, each of the clans and sub-clans stand in a definite relation to the four cardinal points.[3] Lastly, among the Ponka[4] we can go still further. As among the previous tribes, the circle formed by the tribe is divided into two equal halves corresponding to the two moieties. On the other hand, each moiety comprises four clans, but these are quite naturally reducible to two pairs; for the same characteristic element is attributed to two clans at the same time. From this results the following disposition of people and things. The circle is divided into four parts. In the first, to the left of the entrance, are

[1] In order to appreciate how little the orientation of the clans is determined by relation to the cardinal points, it is enough to realize that it changes completely according to whether the route followed by the tribe goes from north to south, or west to east, or the other way. Dorsey and MacGee are rash, therefore, in relating this Omaha system, to the extent that they have done, to a complete classification of clans and things by regions (1894, pp. 522 ff.; MacGee 1894, p. 204).

[2] Dorsey 1896, p. 233; cf. p. 214.

[3] In the ceremony of circumambulation around the cardinal points, the point of departure varies according to clan (Dorsey 1896, p. 380).

[4] Dorsey 1896, p. 220; 1894, p. 523. This tribe has sub-totems of some importance.

two fire clans (or thunder clans); in the part situated at the back, two wind clans; in the first to the right, two water clans; and beyond, two earth clans. Each of the four elements is thus localized exactly in one of the four arcs of the total circumference. Given this, it is only necessary for the axis of this circumference to coincide with one of the two axes of the compass for clans and things to be oriented with relation to the cardinal points. And we know that in these tribes the entrance to the camp generally faces west.[1]

But this orientation (which is partly hypothetical) remains indirect. The secondary groups of the tribe, together with everything subject to them, are situated in quarters of the camp which are more or less clearly oriented; but in not one of these cases is it reported that the clan stands in a particular relationship to any part of space in general. It is still a question of tribal space alone; so we continue to be fairly far from the Zuñi situation.[2] To get close to this we have to leave America and return to Australia. We shall find

[1] Among the Omaha, where the same distribution of clans and things is found, the entrance is not to the west (Dorsey 1894, p. 523). [Durkheim and Mauss refer to the Winnebago, and write that the entrance is to the 'west'. The diagram to which their statement apparently relates actually concerns the 'Omaha, Iowa and cognate tribes'. Dorsey does not say to which compass point the entrance lies, though evidences in the succeeding pages indicate the north-west. The reference in the original to a work by Foster is erroneous.—R. N.] But this different orientation of the entrance does not change the general aspect of the camp.—The same disposition is found, not only in the general assembly of the tribe, but in the individual assemblies of the clans, or at least certain clans. This is notably the case with the Chatada clan. In the circle which it forms when it gathers, earth, fire, wind, and water are situated in exactly the same manner in four different sectors (Dorsey 1894, p. 523).

[2] There is, however, one Sioux tribe in which we do find things classed by quarters as among the Zuñi, viz. the Dakota. But the clans have disappeared among this people, and consequently the classification by clans (Dorsey 1894, pp. 522, 529, 530, 532, 537; cf. Riggs 1885, p. 61). This prevents us taking account of them in our demonstration. The Dakota classification singularly resembles the Chinese classification which we shall examine shortly.

in an Australian tribe a part of what we lack among the
Sioux, which is a new and particularly decisive proof that
the differences between what we have so far called the
American system and the Australian system are not matters
of simply local causes and are in no way irreducible.

This tribe is the Wotjobaluk, which we have already
examined. It is true that Howitt, to whom we owe our
information, does not say that the cardinal points play any
part in the classification of things; and we have no reason
at all to suspect the exactitude of his observations on this
point. But as far as the clans are concerned there is no
doubt at all; each of them is connected with a particular
spatial region which is entirely its own. And this time it is
not a question of a quarter of the camp, but of a delimited
portion of the horizon in general. Each clan can thus be
situated on the compass-card. The relation between the
clan and its spatial region is so intimate that its members
must be buried in the direction thus determined.[1] For
example, a Wartwut, hot wind,[2] is buried 'with the head a
little to the west of north, that is, in the direction from
which the hot wind blows in their country'. The sun-
people are buried in the direction of the sunrise, and so on
for the others.[3]

This division into spatial regions is so closely linked to
the essence of the social organization of this tribe that Howitt
sees it as a 'mechanical method used by the Wotjobaluk to
preserve and explain a record of their classes and totems,
and of their relation to those and to each other'.[4] Two

[1] Howitt 1887, p. 31; 1889a, p. 62.

[2] The word Wartwut means at the same time north and wind from
the north, or hot wind (Howitt 1889a, p. 62, fn. 2). [Durkheim and
Mauss have 'wind from the north-north-west'. Howitt actually says
'North = Wartwut, by which name the hot-wind is also known'.—
R. N.]

[3] Howitt 1887, p. 31.

[4] Howitt 1889a, pp. 62 ff. What follows is a résumé of the text.

clans cannot be related without being *ipso facto* connected with two neighbouring regions in space. This is shown in the Figure, which Howitt constructed according to statements made by a highly intelligent native.[1] The latter, in order to describe the organization of the tribe, began by laying a stick pointing exactly to the east, since Ngaui, the sun, is the principal totem and all the others are determined

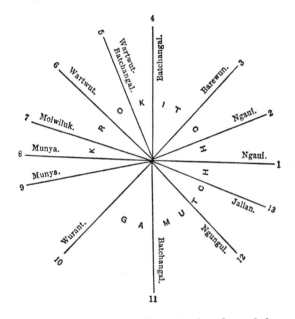

in relation to it. In other words, it is the clan of the sun and the east–west orientation which must have provided the general orientation of the two moieties Krokitch and Gamutch, the former being situated above the east–west line, the other below. In fact, it can be seen from the

[1] The following, so far as can be established, are the translations of the native terms designating the clans: 1 and 2 (Ngaui) mean 'sun'; 3 (Barewun), 'cave' (?); 4 and 11 (Batchangal), 'pelican'; 5 (Wartwut-Batchangal), 'hot wind-pelican'; 6 (Wartwut), 'hot wind'; 7 (Moi), 'carpet snake'; 8 and 9 (Munya), 'kangaroo' (?); 10 (Wurant), 'black cockatoo'; 12 (Ngungul), 'sea'; 13 (Jallan), 'death-adder'.

Figure that the Gamutch moiety is situated entirely to the south, the other almost entirely to the north. A single Krokitch clan, No. 9, crosses over the east–west line, and we have every reason to believe that this anomaly is due to an error of observation or to a more or less late alteration in the original system.[1] This would give us a moiety of the north and a moiety of the south, completely analogous to such as we have seen in other societies. The north–south line is fixed very exactly in the northern part by the pelican clan of Krokitch moiety, and in the southern part by the clan bearing the same name in Gamutch moiety. There are thus four sectors in which the other clans are located. As among the Omaha, the order in which they are arranged expresses relations of kinship between their totems. The spaces separating the clans bear the names of their primary clans, of which the others are segments. Thus clans 1 and 2 are described as 'men of the sun'; the space between them is 'wholly' of the white cockatoo. * Since the white cockatoo is a synonym of the sun, as we have already shown, we may say that the whole of the sector between the east and the north is that of the sun. Similarly, the clans from 4 to 9, i.e. those going from north to west, are all segments of the pelican clan of the first moiety. It may be seen with what regularity, then, that things are oriented.

To sum up, we have reason to think that classification by clans and totems is the older, not only where the two types of classification co-exist, as among the Zuñi, but by review-

---

[1] In fact, Howitt himself mentions that his informant had hesitations on this point. Moreover, this clan is in reality the same as clan 8 and is distinguished from it only by its mortuary totems.

* [This is the situation as described by Howitt (1889a, pp. 62–3). Durkheim and Mauss misrender his report as saying further that the space between clans 1 and 2 belongs to the sun, and that clans 2 and 4 are themselves of the white cockatoo.—R. N.]

ing different societies we have been able to follow the course by which the second system developed from the first and was added to it.

In societies with a totemic organization, it is a general rule that secondary groups of the tribe—moieties, clans, sub-clans—are spatially disposed according to their relations of kinship and the similarities or differences of their social functions. Because the two moieties have distinct personalities, because each has a different role in the life of the tribe, they are spatially opposed; one is established on one side, the other on the other side; one is oriented in one direction, the other in the opposite. Within each moiety, according as the clans are neighbours, or are separated from each other, the things connected respectively with them are also more closely related or are alien to each other. The existence of this rule is quite apparent in the societies of which we have spoken. We have seen, in fact, how among the Zuñi each clan within the pueblo is oriented in the direction of the region assigned to it; how among the Sioux the two moieties, possessing functions as opposite as may well be, are situated one to the left, the other to the right, one to the east, the other to the west. But identical or similar facts are found in many other tribes. This double opposition of moieties is reported, with regard both to function and to localization, among the Iroquois,[1] the Wyandot,[2] the disintegrated Seminole tribe of Florida,[3] the Tlingit, and the Loucheux or Déné Dindjé, the most northern, the most bastardized, but also the most primitive of Indians.[4] The relative localization of moieties and clans is no less rigorously determined in Melanesia. It suffices to recall the fact, already mentioned, that these tribes are divided into

[1] Morgan 1877, pp. 88, 94–5; 1851, pp. 294 ff.; Smith 1883, p. 114.
[2] Powell 1883, p. 44.   [3] MacCauley 1887, pp. 507–9.
[4] Petitot 1887, pp. 15 and 20. Among the Loucheux there is one phratry of the right, one of the left, and one of the middle.

moieties of sea and land, one camping downwind, the other upwind.[1] In many Melanesian societies this bipartite division is in fact all that remains of the former organization.[2] The same phenomena of localization have been reported, on many occasions, from Australia. Even though the members of each moiety are dispersed over a multitude of local groups, within each group they are opposed to each other in their camps.[3] But these dispositions, and the orientation resulting from them, are apparent above all in the gatherings of the entire tribe. This is particularly the case among the Arunta. Moreover, we find among them the idea of a special orientation, a mythical direction assigned to each clan. The water clan belongs to a region thought to be that of water.[4] The dead are oriented in the direction of the mythical camp thought to have been lived in by the legendary ancestors, the Alcheringa. The direction of the camp of the mother's mythical ancestors is taken into account at certain religious ceremonies (nose-piercing, extraction of the upper incisor).[5] Among the Kulin, and in the entire group of tribes inhabiting the coast of New South Wales, the clans are placed in the tribal assembly according to the point on the horizon from which they came.[6]

This seen, we easily understand how a classification by

[1] See above, p. 28.

[2] Pfeil 1899, p. 28.

[3] Spencer and Gillen 1899, pp. 32, 70, 277, 287, 324, 501.

[4] Spencer and Gillen 1899, p. 189.

[5] Spencer and Gillen 1899, p. 496. This is clearly a case of either an incipient or a defunct localization of clans. We think it is more likely to be the latter, of which we see traces. If it is admitted that the clans were divided between the moieties, the demonstration of which has already been attempted (Émile Durkheim, 'Sur le totémisme', *Année Sociologique*, vol. V, 1902, pp. 82–121), then as the moieties were localized so must the clans have been as well.

[6] Howitt 1884c, pp. 441, 442. Similarly, among the Kamilaroi (Mathews 1895, p. 414; 1896, pp. 322, 326).

orients was established. Things were first of all classified by clans and totems. But this strict localization of clans which we have just described necessarily brought with it a corresponding localization of the things attributed to the clans. From the moment that the wolf people, for example, belong to a particular quarter of the camp, the same necessarily applies to the things of all sorts which are classified under this same totem. Consequently, the camp has only to be oriented in a fixed way and all its parts are immediately oriented, together with everything, things and people, that they comprise. In other words, all things in nature are henceforth thought of as standing in fixed relationships to equally fixed regions in space. Certainly, it is only tribal space which is divided and shared in this way. But just as for the primitive the tribe constitutes all humanity, and as the founding ancestor of the tribe is the father and creator of men, so also the idea of the camp is identified with that of the world.[1] The camp is the centre of the universe, and the whole universe is concentrated within it. Cosmic space and tribal space are thus only very imperfectly distinguished, and the mind passes from one to the other without difficulty, almost without being aware of doing so. And in this way things are connected with particular quarters. However, as long as the organization into moieties and clans remained strong, the classification by clans was preponderant; it was through the totems that things were attached to regions. We have seen that this is still the case among the Zuñi, at least for certain things. But if the totemic groups, so curiously hierarchized, vanish or are replaced by local groupings which are simply juxtaposed to

[1] Traces of these ideas are found also in Rome: *mundus* means both the world and the place where the *comitia* gathered. The identification of the tribe (or city) with humanity is thus not due simply to the exaltation of national pride, but to a system of ideas which see in the tribe the *microcosm of the universe*.

each other, then concordantly a classification by quarters is the only possible one.[1]

Thus the two types of classification which we have just studied merely express under different aspects the very societies within which they were elaborated; one was modelled on the jural and religious organization of the tribe, the other on its morphological organization. When it was a matter of establishing ties of kinship between things, and of constituting more and more vast families of creatures and phenomena, this was done with the aid of ideas supplied by the family, the clan, and the moiety, and the totemic myths were taken as starting point. When it was a matter of establishing relations between spatial regions, it was the spatial relations which people maintained within their society that served as starting point. In one case, the framework was furnished by the clan itself, in the other by the material mark made on the ground by the clan. But both forms are of social origin.

[1] In this case, all that survives of the former system is the attribution of certain powers to local groups. Thus among the Kurnai each local group is master of a certain wind which is thought to come from its side.

Chapter Four

CHINA

WE HAVE NOW to describe, at least in its principles, a final type of classification which presents all the essential characteristics of the preceding ones except that it is independent, and has been for as long as it has been known, of any social organization. The best case of the kind, the most remarkable and the most instructive, is the astronomical, astrological, geomantic and horoscopic divinatory system of the Chinese. This system has behind it a history going back to the most distant times; for it is certainly older than the first authentic and dated documents which have survived in China.[1] By the first centuries of our era it was already fully developed. On the other hand, while we shall study it by preference in China, this is not because it is peculiar to that country; it is found everywhere in the Far East.[2] The Siamese, Cambodians, Tibetans and the Mongols all know it and use it. For all of these peoples it expresses the 'Tao', i.e. nature. It is fundamental to the entire philosophy and cult commonly known as Taoism.[3] It governs all details of life among the most immense population that humanity has ever known.

The very importance of this system permits us to do no

[1] De Groot 1892–1910, p. 319; cf. pp. 982 ff.
[2] De Groot 1892–1910, p. 989.
[3] De Groot 1892–1910, p. 989.

more than to trace its main features. We shall confine ourselves to describing only what is strictly necessary in order
to show how it agrees, in general principles, with those
which we have so far described.

The system is itself composed of a number of intermingled
systems.

One of the most essential principles on which it rests is a
division of space according to the four cardinal points. An
animal presides over each of these four regions and gives its
name to it. More exactly, the animal is identified with its
region: the azure dragon is the east, the red bird is the
south, the white tiger is the west, the black tortoise is the
north. Each region has the colour of its animal, and according to varying conditions which we cannot recount here it is
favourable or unfavourable. Moreover, symbolic creatures
which are thus set in charge over space govern the earth as
well as the sky. Thus a hill or geographic configuration
which looks like a tiger belongs to the tiger and to the west;
if it resembles a dragon, it belongs to the dragon and to the
east. Consequently, a site will be considered favourable if
the things surrounding it bear aspects conforming to their
orientation; for example, if those to the west are of the
tiger and those to the east are of the dragon.[1]

But the space contained between each cardinal point is
itself divided into two parts: this results in a total of eight
divisions[2] corresponding to eight compass points. These
compass points, in their turn, are closely connected with
eight powers, represented by eight trigrams occupying the

[1] The matter, moreover, is even more complicated; seven constellations are divided among the four regions, giving 28 Chinese asterisms.
(It is well known that many scholars attribute a Chinese origin to
asterisms in the entire Orient.) Astral, terrestrial, and atmospheric
influences all combine in this so-called *fung-shui*, or 'wind and water',
system. On this system, see De Groot (1892–1910, vol. III, book I,
chap. XII, and references cited above).

[2] De Groot 1892–1910, p. 960.

centre of the divinatory compass. These eight powers are, firstly, those at the two extremities (the first and the eighth), the two opposed substances of earth and sky; and between these are situated the six other powers, viz. (1) mists, clouds, emanations, etc.; (2) fire, heat, sun, light, lightning; (3) thunder; (4) wind and wood; (5) water, rivers, lakes and the sea; and (6) mountains.

Here we have then a certain number of fundamental elements, classed with different points of the compass. Now to each of these is attached a whole collection of things: *khien*, the sky, pure principle of light, male, etc., is placed to the south.[1] It 'represents' immobility and force, the head, the heavenly sphere, a father, a prince, roundness, jade, metal, ice, red, a good horse, an old horse, a thin horse,* the fruit of trees, etc. In other words, the sky connotes these different sorts of things in the way that, among ourselves, the genus connotes the species which it includes. *Khwun*, feminine principle, principle of the earth and darkness, is to the north; it covers docility, cattle, the belly, mother earth, cloth, cauldrons, multitude, black, large carts, etc. 'Sun' means penetration; under it are subsumed wind, wood, length, height, fowls, thighs, eldest daughter, forward and backward movements, any gain of three hundred per cent., etc.† We restrict ourselves to these few examples. The list of species of creatures, events, attributes, substances and accidents thus classified under the eight powers is truly in-

[1] See chapter XV of the *Yî-King*, in Legge's translation (1882). We follow the table drawn up by de Groot (p. 964). Naturally, these classifications lack anything resembling Greek or European logic. Contradictions, deviations, and overlappings abound in them. They are all the more interesting in our eyes on account of these.

* [Durkheim and Mauss have 'a fat horse' (*un gros cheval*). They follow this with 'a sabre' (*un bancale*), but this is not in the source; at this place de Groot has 'a piebald horse'.—R. N.]

† [For 'three hundred' Durkheim and Mauss have '3'. Here and elsewhere in this paragraph the order in which things are listed is considerably different from their order in the source.—R. N.]

finite. It covers the whole world in the fashion of a gnosis or cabbala. The classical writers and their imitators abandon themselves, with an inexhaustible verve, to endless speculations on this theme.

In addition to the classification by eight powers there is another which distributes things under five elements, earth, water, wood, metal, and fire. It has been remarked, though, that the former is not irreducible to the latter; if, that is, the mountains are eliminated, and if on the other hand mists are merged with water, and thunder with fire, the two divisions coincide exactly.

Whatever is made of the question whether these two classifications are derived one from the other or superimposed one on the other, the elements play the same part as the powers. Not only is everything connected with them, according to the substances of which they are composed or according to their forms, but so also are historical events, contours of the ground, etc.[1] The planets themselves are attributed to them: Venus is the star of metal, Mars the star of fire, etc. On the other hand, this classification is linked to the system as whole by the fact that each of the elements is localized in one fundamental division. It was enough to place earth at the centre of the universe, as was reasonable enough, to be able to apportion the elements to the four spatial regions. Consequently, they too, like the regions, are good or bad, powerful or weak, generators or created.

We shall not follow Chinese philosophy in its thousands of traditional elaborations. In order to adapt the basic principles of the system to the facts, the divisions and subdivisions of regions and things were ceaselessly multiplied and complicated. There was no fear, even, of the most obvious contradictions. For example, it was found possible

[1] De Groot 1892–1910, p. 956.

to see earth as situated alternatively to the north, to the north-east, and at the centre. The fact is that this classification was intended above all to regulate the conduct of men; and it was able to do so, avoiding the contradictions of experience, thanks to this very complexity.

There remains to be explained, however, a last complication of the Chinese system: like space, and like things and events, time itself forms part of it. The four seasons correspond to the four regions. Moreover, each of these regions is sub-divided into six parts, and the twenty-four sub-divisions give naturally the twenty-four seasons of the Chinese year.[1] There is nothing surprising in this concordance. In all the systems of thought that we have spoken of above, the importance of the seasons is parallel to that of space.[2] As soon as an orientation is made, the seasons are necessarily related to the cardinal points, winter with the north, summer with the south, and so on. But the distinction of seasons is only a first step in the reckoning of time. In order to be complete, this supposes in addition a division into cycles, years, days, and hours which permits the measurement of any period of time, large or small. The Chinese arrived at this result by the following procedure. They constructed two cycles, one of twelve divisions and the other of ten; each of these divisions has its own name and nature, so that any moment of time is represented by binomial characters taken from the two different cycles.[3] These two cycles are employed concurrently, for both years and days, months and hours, and a fairly exact measurement is thus arrived at. Their combination forms, consequently, a sexagesimal cycle,[4] since after five revolutions of the cycle of twelve, and six revolutions of the cycle of ten, the

[1] De Groot 1892–1910, p. 968. [2] See above, p. 43.

[3] See de Groot 1892–1910, pp. 966, 973. In the most ancient classics they are called the ten mothers and the twelve children.

[4] The duodecimal and sexagesimal divisions served as bases for the

same binomial characters return exactly to qualify the same period of time. Just like the seasons, the two cycles with their divisions are linked to the points of the compass,[1] and, through the intermediary of the four cardinal points, to the five elements; and it is thus that the Chinese arrived at the notion, an extraordinary one to our current ideas, of a non-homogeneous time, symbolized by the elements, the cardinal points, colours, and things of every kind subsumed under them, and over the different parts of which the most various influences predominate.[2]

This is not all. The twelve years of the sexagenary cycle are further connected with twelve animals arranged in the following order: rat, cow, tiger, hare, dragon, snake, horse, goat, monkey, hen, dog, pig.[3] These twelve animals are divided three by three between the four cardinal points and in this way too this division of time[4] is linked to the general system. Thus, say texts dating from the beginning of our era, 'a *tsze* year has the rat as animal, and belongs to north and water; a *wu* year belongs to fire, i.e. to the south, and its animal is the horse', etc.[*] Subsumed under the elements,[5] the years are likewise subsumed under the

---

[1] De Groot 1892–1910, p. 967.   [2] De Groot 1892–1910, pp. 968–88.
[3] De Groot 1892–1910, pp. 44, 987.

[4] We cannot refrain from thinking that the cycle of twelve divisions, and the twelve years represented by animals, were originally nothing but one and the same division of time, one esoteric, the other exoteric. One text calls them 'the two dozens which belong to each other'; which appears to indicate that they were only one dozen, differently symbolized.

[*] [De Groot's text reads: 'Tsze is identical with water, and its animal is the rat; and Wu appertains to fire, and its animal is the horse'. The references to cardinal points are interpolations.—R. N.]

[5] Here again, the elements are no more than four: the earth ceases to be an element and becomes a first principle. This arrangement was necessary in order to permit the establishment of an arithmetical relationship between the elements and the twelve animals. Contradictions are infinite.

Chinese computation of the celestial circle and for the division of the divinatory compass.

regions, which themselves are represented by animals. We clearly have to do with a multitude of interlaced classifications which, in spite of their contradictions, grasp reality closely enough to provide a fairly useful guide to action.[1]

This classification of regions, seasons, things, and animal species dominates the whole of Chinese life. It is the very principle of the famous doctrine of *fung-shui*, and through this it determines the orientation of buildings, the foundation of towns and houses, the siting of tombs and cemeteries; if certain tasks are undertaken here and others there, if certain affairs are conducted at such and such a time, this is due to reasons based on this traditional systematization. And these reasons are not taken only from geomancy; they are also derived from considerations concerning hours, days, months, and years: a certain direction which is favourable at one time becomes unfavourable at another. Forces agree or discord according to season. Thus not only is everything heterogeneous in time, as in space, but the heterogeneous parts of which these two settings are composed correspond, or are opposed, and are arranged, in one system. And all these infinitely numerous elements are combined to determine the genus and the species of things in nature, the direction of movement of forces, and acts which must be performed, thus giving the impression of a philosophy which is at once subtle and naive, rudimentary and refined. Here we have, then, a highly typical case in which collective thought has worked in a reflective and learned way on themes which are clearly primitive.

Indeed, though we have no means of establishing an historical link between the Chinese system and the types of

[1] Williams 1899, vol. II, pp. 69 ff. Williams reduces the denary cycle to the five elements, each couple of the decimal division corresponding to one element. It is very possible, too, that the denary division was part of an orientation into five regions, and the duodenary division part of an orientation into four cardinal points.

classification that we studied earlier, it is impossible not to remark that it is based on the same principles as they are. The classification of things under eight headings, the eight powers, actually gives a division of the universe into eight families which is comparable, save for the fact that the notion of clan is absent, to the Australian classifications. Also, we have found at the basis of the system, as among the Zuñi, a completely analogous division of space into fundamental regions. These regions are likewise connected with the elements, compass points, and the seasons. As among the Zuñi, again, each region has its own colour and is placed under the preponderant influence of a certain animal, which symbolizes at once the elements, powers, and moments of time. It is true that we have no means of proving decisively that these animals were ever totems. Whatever importance clans have retained in China, and even though they still possess the distinctive feature of strictly totemic clans, viz. exogamy,[1] it does not seem that they formerly bore the names which are used to designate regions or hours. But it is none the less curious that in Siam, according to a contemporary author, there should be a prohibition on marriage between persons of the same year and the same animal, even though this year may belong to two different duodecades;[2] i.e. that the relationship between the individuals

[1] Williams 1899, vol. I, p. 792.

[2] Young 1898, p. 92. [Young actually says only that 'persons born in certain years should not marry each other, as any union between them would only be fruitful of endless discord. Thus a person born in the "year of the Dog" might lead a life of never ending discord with one born in the "year of the Rat".' There is no mention of a prohibition on marriage between persons of the same year and animal, of whatever duodecades.—R. N.] Other authors mention only the consultation of diviners and the inspection of cycles. See Pallegoix 1854, vol. I, p. 253; 1896, p. ii; Chevillard 1889, p. 252, cf. p. 154; La Loubère 1714, vol. I, p. 156, vol. II, p. 62.

This cycle seems to have a rather complicated history. In Cambodia the cycle is employed as in China (Moura 1878, p. 15). But neither

and the animal with which they are connected has exactly the same effect on conjugal relations as that in which, in other societies, they stand to their totems. Besides, we know that in China the horoscope, the examination of the eight characters, plays a considerable part in the consultation of diviners preliminary to any matrimonial interview.[1] It is true that none of the authors we have consulted mentions a marriage between two individuals of the same year, or two years of the same name, as legally forbidden. However, it is probable that such a marriage is regarded as particularly inauspicious. In any case, although we do not find in China this sort of exogamy between people born under the same animal, there nevertheless exists between them, from another point of view, a relationship which is quasi-familial. Doolittle in fact tells us that every individual is thought to belong to a particular animal,[2] and those belonging to the same animal may not attend each other's funerals.[3]

China is not the only civilized country where we find at least traces of a classification recalling those observed in simpler societies.

[1] Doolittle 1876, vol. I, pp. 66 and 69.

[2] Doolittle 1876, vol. II, p. 341.

[3] Doolittle 1876, vol. II, p. 342. Cf. de Groot 1892–1910, vol. I, book I, part 1, p. 106, where the same fact appears to be mentioned in a different form. [This is not the sense of Doolittle's account. A fortune-teller computes that 'a certain animal' is to be feared or avoided at the time an event is to take place. 'This means simply that those persons who were born during the year denoting the specified animal should not be present when the event referred to is to transpire, as a house-raising, or the putting of a corpse into the coffin, or the celebration of a certain marriage, etc.' It is not said that those who should not attend a funeral are of the same animal as the deceased. Moreover, the place in

authors nor codes say anything about matrimonial prohibitions connected with the cycle (Leclère 1898). It is probable, therefore, that it was quite simply a belief originating exclusively in divination, and all the more popular in that Chinese divination is used more in these societies.

First of all, we have just seen that the Chinese classification was essentially an instrument of divination. Now the divinatory methods of Greece are remarkably similar to the Chinese, and the similarities denote procedures of the same nature in the way fundamental ideas are classified.[1] The assignment of elements and metals to the planets is a Greek, perhaps Chaldaean, fact, as much as a Chinese. Mars is fire, Saturn is water, etc.[2] The relation between certain sorts of events and certain planets, the simultaneous apprehension of space and time, the particular correspondence of a certain region with a certain time of the year and with a certain kind of undertaking, are found equally in both these different societies.[3] A still more curious coincidence is that which allows a relationship to be established between Chinese and Greek astrology and physiognomy, and perhaps with the Egyptian. The Greek theory of zodiacal and planetary melothesia, which is thought to be of Egyptian origin,[4] is intended to establish strict correspondences between certain parts of the body and certain positions of the stars, certain orientations, and certain events. Now in China also

[1] It has even been conjectured whether there might not have been borrowing from one of these peoples by the other.

[2] Bouché-Leclercq 1899, pp. 311 ff., 316.

[3] Epicurus criticizes precisely prognostications based on (celestial?) animals as being based on the hypothesis of the coincidence of time, directions and events created by the divinity (Usener 1887, p. 55, l. 13).

[4] Bouché-Leclercq 1899, pp. 319, 76 ff. Cf. Ebers 1901.

---

de Groot to which Durkheim and Mauss evidently refer lends no support to their inference. The text reads: 'No transaction of importance can by any means bring good luck to the person who performs it or acts a leading part in it, if it is performed on a day or hour the cyclical characters of which stand, in the [divinatory] circle, opposite to characters occurring in the horoscope of that person. . . . Perfection is reached when the calculations are applied to years and months also. But then . . . the choice is so greatly reduced as to render it almost impossible for the day-professor to arrive at any decision' (p. 104). The conclusion from these particulars must clearly be that it is not simply, if at all, persons of the same animal as the deceased who must not attend the funeral rites.—R. N.]

there exists a famous doctrine based on the same principle. Each element is related to a cardinal point, a constellation, and a particular colour, and these different groups of things are thought to correspond, in turn, to diverse kinds of organs, inhabited by various souls, to emotions, and to different parts whose reunion forms 'the natural character'. Thus, *yang*, the male principle of light and sky, has the liver in the viscera, the bladder as mansion, and the ears and sphincters among the orifices.[1] This theory, the generality of which is apparent, is not of mere curiosity-value; it implies a certain way of conceiving things. By it, the universe is in fact referred to the individual; things are expressed by it, in a sense, as functions of the living organism; this is really a theory of the microcosm.

There is nothing more natural, moreover, than the relation thus expressed between divination and the classification of things. Every divinatory rite, however simple it may be, rests on a pre-existing sympathy between certain beings, and on a traditionally admitted kinship between a certain sign and a certain future event. Further, a divinatory rite is generally not isolated; it is part of an organized whole. The science of the diviners, therefore, does not form isolated groups of things, but binds these groups to each other. At the basis of a system of divination there is thus, at least implicitly, a system of classification.

But it is above all in myths that we see the appearance, in an almost ostensible manner, of methods of classification entirely analogous to those of the Australians or North American Indians. Every mythology is fundamentally a classification, but one which borrows its principles from

[1] According to Pan-ku, an author of the second century, basing himself on much more ancient sources (de Groot 1892–1910, vol. IV, pp. 13 ff.). [The cryptic example given is an exiguous and mangled indication of a complex and systematic exposition on pp. 13–25 of de Groot.—R. N.]

religious beliefs, not from scientific ideas. Highly organized pantheons divide up all nature, just as elsewhere the clans divide the universe. Thus India divides things, as well as their gods, between the three worlds of the sky, the atmosphere, and the earth, just as the Chinese class everything according to the two fundamental principles of *yang* and *yin*. To attribute certain things in nature to a god amounts to the same thing as to group them under the same generic rubric, or to place them in the same class; and the genealogies and identifications relating divinities to each other imply relations of co-ordination or subordination between the classes of things represented by these divinities. When Zeus, father of men and the gods, is said to have given birth to Athena, the warrior-goddess, goddess of intelligence, mistress of the owl, etc., this really means that two groups of images are linked and classified in relation to each other. Every god has his doubles, who are other forms of himself, though they have other functions; hence, different powers, and the things over which these powers are exercised, are attached to a central or predominant notion, as is the species to the genus or a secondary variety to the principal species. It is thus that to Poseidon,[1] the river god, are attached other and paler personalities, agrarian gods (Aphareus, Aloeus, the farmer, the thresher), horse gods (Actor, Elatos, Hippocoon, etc.), and a vegetation god (Phutalmios).

These classifications are such essential elements of developed mythologies that they have played an important part in the evolution of religious thought; they have facilitated the reduction of a multiplicity of gods to one, and consequently they have prepared the way for monotheism. The 'henotheism' [2] which characterizes Brahmanic mythology,

---

[1] Usener 1898, p. 357.

[2] The word is Max Müller's, but he is mistaken in applying it to primitive forms of Brahmanism.

at least after it has reached a certain stage of development, actually consists in a tendency to reduce more and more gods into each other, to the extent that each ends up by possessing the attributes of all the others and even their names. The pantheism of pre-Buddhist India is, from a certain point of view, an unstable classification in which the genus easily becomes a species, and *vice versa*, but which manifests an increasing tendency towards unity; and the same is true of classical Śivaism and Vishnuism.[1] Usener has similarly shown[2] that the progressive systematization of Greek and Roman polytheism was an essential condition for the advent of western monotheism.* Minor local and specialized gods are gradually subsumed under more general headings, the great nature gods, and tend to be absorbed by them. For a time, the idea of what was peculiar to the former remains; the name of the old god coexists with that of the great god, but only as an attribute of the latter; then his existence becomes more and more than of a phantom, until one day only the great gods remain, if not in religious observances, at least in myth. One might almost say that mythological classifications, when they are complete and systematic, when they embrace the universe, announce the end of mythologies properly speaking. Pan, Brahmán, Prajāpati, supreme genera, absolute and pure beings, are mythical figures almost as poor in imagery as the transcendental God of the Christians.

Thence it seems that we approach imperceptibly the abstract and relatively rational types which crown the first philosophical classifications. It is certain that Chinese philosophy, when it is really Taoist, is based on the system of classification that we have described. In Greece, without

[1] Barth 1891, pp. 29, 160 ff.   [2] Usener 1896, pp. 346 ff.

* [Durkheim and Mauss have 'polytheism', but see especially p. 347 of the source cited: 'So ist . . . der polytheismus . . . zu monotheistischer vorstellung hingeführt worden.'—R. N.]

wishing to affirm anything about the historical origin of its doctrines, one cannot but remark that the two principles of Heraclitean Ionism, viz. war and peace, and those of Empedocles, viz. love and strife, divide things between them in the same way as do *yang* and *yin* in the Chinese classification. The relationships established by the Pythagoreans between numbers, elements, sexes, and a certain number of other things are reminiscent of the correspondences of magico-religious origin which we have had occasion to discuss. Also, even in the time of Plato, the world was still conceived as a vast system of classified and hierarchized sympathies.[1]

---

[1] Hindu philosophy abounds in correspondential classifications of things, elements, directions, and hypostases. The main ones are listed, with commentary, in Deussen (1894, vol. I, part 2, pp. 85, 89, 95, etc.). A large part of the Upanishads consists in speculations on genealogies and correspondences.

Chapter Five

CONCLUSIONS

PRIMITIVE CLASSIFICATIONS are therefore not singular or exceptional, having no analogy with those employed by more civilized peoples; on the contrary, they seem to be connected, with no break in continuity, to the first scientific classifications. In fact, however different they may be in certain respects from the latter, they nevertheless have all their essential characteristics. First of all, like all sophisticated classifications, they are systems of hierarchized notions. Things are not simply arranged by them in the form of isolated groups, but these groups stand in fixed relationships to each other and together form a single whole. Moreover, these systems, like those of science, have a purely speculative purpose. Their object is not to facilitate action, but to advance understanding, to make intelligible the relations which exist between things. Given certain concepts which are considered to be fundamental, the mind feels the need to connect to them the ideas which it forms about other things. Such classifications are thus intended, above all, to connect ideas, to unify knowledge; as such, they may be said without inexactitude to be scientific, and to constitute a first philosophy of nature.[1] The Australian

[1] As such they are very clearly distinguished from what might be called technological classifications. It is probable that man has always classified, more or less clearly, the things on which he lived, according

does not divide the universe between the totems of his tribe with a view to regulating his conduct or even to justify his practice; it is because, the idea of the totem being cardinal for him, he is under a necessity to place everything else that he knows in relation to it. We may therefore think that the conditions on which these very ancient classifications depend may have played an important part in the genesis of the classificatory function in general.

Now it results from this study that the nature of these conditions is social. Far from it being the case, as Frazer seems to think, that the social relations of men are based on logical relations between things, in reality it is the former which have provided the prototype for the latter. According to him, men were divided into clans by a pre-existing classification of things; but, quite on the contrary, they classified things because they were divided by clans.

We have seen, indeed, how these classifications were modelled on the closest and most fundamental form of social organization. This, however, is not going far enough. Society was not simply a model which classificatory thought followed; it was its own divisions which served as divisions for the system of classification. The first logical categories were social categories; the first classes of things were classes of men, into which these things were integrated. It was because men were grouped, and thought of themselves in the form of groups, that in their ideas they grouped other things, and in the beginning the two modes

---

to the means he used to get them: for example, animals living in the water, or in the air or on the ground. But at first such groups were not connected with each other or systematized. They were divisions, distinctions of ideas, not schemes of classification. Moreover, it is evident that these distinctions are closely linked to practical concerns, of which they merely express certain aspects. It is for this reason that we have not spoken of them in this work, in which we have tried above all to throw some light on the origins of the logical procedure which is the basis of scientific classifications.

of grouping were merged to the point of being indistinct. Moieties were the first genera; clans, the first species. Things were thought to be integral parts of society, and it was their place in society which determined their place in nature. We may even wonder whether the schematic manner in which genera are ordinarily conceived may not have depended in part on the same influences. It is a fact of current observation that the things which they comprise are generally imagined as situated in a sort of ideational milieu, with a more or less clearly delimited spatial circumscription. It is certainly not without cause that concepts and their interrelations have so often been represented by concentric and eccentric circles, interior and exterior to each other, etc. Might it not be that this tendency to imagine purely logical groupings in a form contrasting so much with their true nature originated in the fact that at first they were conceived in the form of social groups occupying, consequently, definite positions in space? And have we not in fact seen this spatial localization of genus and species in a fairly large number of very different societies?

Not only the external form of classes, but also the relations uniting them to each other, are of social origin. It is because human groups fit one into another—the sub-clan into the clan, the clan into the moiety, the moiety into the tribe—that groups of things are ordered in the same way. Their regular diminution in span, from genus to species, species to variety, and so on, comes from the equally diminishing extent presented by social groups as one leaves the largest and oldest and approaches the more recent and the more derivative. And if the totality of things is conceived as a single system, this is because society itself is seen in the same way. It is a whole, or rather it is *the* unique whole to which everything is related. Thus logical hierarchy is only another aspect of social hierarchy, and the unity of knowledge is nothing else than the very unity of the collectivity, extended to the universe.

Furthermore, the ties which unite things of the same group or different groups to each other are themselves conceived as social ties. We recalled in the beginning that the expressions by which we refer to these relations still have a moral significance; but whereas for us they are hardly more than metaphors, originally they meant what they said. Things of the same class were really considered as relatives of the individuals of the same social group, and consequently of each other. They are of 'the same flesh', the same family. Logical relations are thus, in a sense, domestic relations. Sometimes, too, as we have seen, they are comparable at all points with those which exist between a master and an object possessed, between a chief and his subjects. We may even wonder whether the idea of the pre-eminence of genus over species, which is so strange from a positivistic point of view, may not be seen here in its rudimentary form. Just as, for the realist, the general idea dominates the individual, so the clan totem dominates those of the sub-clans and, still more, the personal totems of individuals; and wherever the moiety has retained its original stability it has a sort of primacy over the divisions of which it is composed and the particular things which are included in them. Though he may be essentially Wartwut and partially Moiwiluk, the Wotjobaluk described by Howitt is above all a Krokitch or a Gamutch. Among the Zuñi, the animals symbolizing the six main clans are set in sovereign charge over their respective sub-clans and over creatures of all kinds which are grouped with them.

But if the foregoing has allowed us to understand how the notion of classes, linked to each other in a single system, could have been born, we still do not know what the forces

were which induced men to divide things as they did between the classes. From the fact that the external form of
the classification was furnished by society, it does not
necessarily follow that the way in which the framework
was used is due to reasons of the same origin. *A priori* it is
very possible that motives of a quite different order should
have determined the way in which things were connected and
merged, or else, on the contrary, distinguished and opposed.

The particular conception of logical connexions which we
now have permits us to reject this hypothesis. We have
just seen, in fact, that they are represented in the form of
familial connexions, or as relations of economic or political
subordination; so that the same sentiments which are the
basis of domestic, social, and other kinds of organization
have been effective in this logical division of things also.
The latter are attracted or opposed to each other in the
same way as men are bound by kinship or opposed in the
vendetta. They are merged as members of the same family
are merged by common sentiment. That some are subordinate to others is analogous in every respect to the fact
that an object possessed appears inferior to its owner, and
likewise the subject to his master. It is thus states of the
collective mind (*âme*) which gave birth to these groupings,
and these states moreover are manifestly affective. There
are sentimental affinities between things as between individuals, and they are classed according to these affinities.

We thus arrive at this conclusion: it is possible to classify
other things than concepts, and otherwise than in accordance with the laws of pure understanding. For in order
for it to be possible for ideas to be systematically arranged
for reasons of sentiment, it is necessary that they should
not be pure ideas, but that they should themselves be products of sentiment. And in fact, for those who are called
primitives, a species of things is not a simple object of

knowledge but corresponds above all to a certain sentimental attitude. All kinds of affective elements combine in the representation made of it. Religious emotions, notably, not only give it a special tinge, but attribute to it the most essential properties of which it is constituted. Things are above all sacred or profane, pure or impure, friends or enemies, favourable or unfavourable;[1] i.e. their most fundamental characteristics are only expressions of the way in which they affect social sensibility. The differences and resemblances which determine the fashion in which they are grouped are more affective than intellectual. This is how it happens that things change their nature, in a way, from society to society; it is because they affect the sentiments of groups differently. What is conceived in one as perfectly homogeneous is represented elsewhere as essentially heterogeneous. For us, space is formed of similar parts which are substitutable one for the other. We have seen, however, that for many peoples it is profoundly differentiated according to regions. This is because each region has its own affective value. Under the influence of diverse sentiments, it is connected with a special religious principle, and consequently it is endowed with virtues *sui generis* which distinguish it from all others. And it is this emotional value of notions which plays the preponderant part in the manner in which ideas are connected or separated. It is the dominant characteristic in classification.

It has quite often been said that man began to conceive things by relating them to himself. The above allows us to see more precisely what this anthropocentrism, which might better be called *sociocentrism*, consists of. The centre of the first schemes of nature is not the individual; it is

---

[1] For the adherent of many cults, even now, foodstuffs are classified first of all into two main classes, fat and lean, and we know to what extent this classification is subjective.

society.[1] It is this that is objectified, not man. Nothing shows this more clearly than the way in which the Sioux retain the whole universe, in a way, within the limits of tribal space; and we have seen how universal space itself is nothing else than the site occupied by the tribe, only indefinitely extended beyond its real limits. It is by virtue of the same mental disposition that so many peoples have placed the centre of the world, 'the navel of the earth', in their own political or religious capital,[2] i.e. at the place which is the centre of their moral life. Similarly, but in another order of ideas, the creative force of the universe and everything in it was first conceived as a mythical ancestor, the generator of the society.

This is how it is that the idea of a logical classification was so hard to form, as we showed at the beginning of this work. It is because a logical classification is a classification of concepts. Now a concept is the notion of a clearly determined group of things; its limits may be marked precisely. Emotion, on the contrary, is something essentially fluid and inconsistent. Its contagious influence spreads far beyond its point of origin, extending to everything about it, so that it is not possible to say where its power of propagation ends. States of an emotional nature necessarily possess the same characteristic. It is not possible to say where they begin or where they end; they lose themselves in each other, and mingle their properties in such a way that they cannot be rigorously categorized. From another point of view, in order to be able to mark out the limits of a class, it is necessary to have analysed the characteristics by which the things assembled in this class are recognized and by which

[1] De la Grasserie has developed ideas fairly similar to our own, though rather obscurely and above all without evidence (1899, chap. III).

[2] Something understandable enough for the Romans and even the Zuñi, but less so for the inhabitants of Easter Island, called Te Pito-te Henua (navel of the earth); but the idea is perfectly natural everywhere.

they are distinguished. Now emotion is naturally refractory to analysis, or at least lends itself uneasily to it, because it is too complex. Above all when it has a collective origin it defies critical and rational examination. The pressure exerted by the group on each of its members does not permit individuals to judge freely the notions which society itself has elaborated and in which it has placed something of its personality. Such constructs are sacred for individuals. Thus the history of scientific classification is, in the last analysis, the history of the stages by which this element of social affectivity has progressively weakened, leaving more and more room for the reflective thought of individuals. But it is not the case that these remote influences which we have just studied have ceased to be felt today. They have left behind them an effect which survives and which is always present; it is the very cadre of all classification, it is the ensemble of mental habits by virtue of which we conceive things and facts in the form of co-ordinated or hierarchized groups.

This example shows what light sociology throws on the genesis, and consequently the functioning, of logical operations. What we have tried to do for classification might equally be attempted for the other functions or fundamental notions of the understanding. We have already had occasion to mention, in passing, how even ideas so abstract as those of time and space are, at each point in their history, closely connected with the corresponding social organization. The same method could help us likewise to understand the manner in which the ideas of cause, substance, and the different modes of reasoning, etc. were formed. As soon as they are posed in sociological terms, all these questions, so long debated by metaphysicians and psychologists, will at last be liberated from the tautologies in which they have languished. At least, this is a new way which deserves to be tried.

BIBLIOGRAPHY

Abbreviations employed:

ARBE   *Annual Report of the Bureau of Ethnology.* Washington.

JAI   *Journal of the Anthropological Institute of Great Britain and Ireland.* London.

BARTH, A. 1891. *The Religions of India.* London.

BASTIAN, ADOLPH. 1887. *Die Welt in ihren Spiegelungen. . . .* Berlin.

   1888. *Allerlei aus Volks- und Menschenkunde.* Berlin.

   1892. *Ideale Welten.* Berlin.

BOUCHÉ-LECLERCQ, A. 1899. *L'Astrologie grècque.* Paris.

BUCKLAND, A. W. 1893. 'Points of contact between Old World customs and the Navaho myth entitled "The Mountain Chant" '. *JAI*, 22, pp. 346–55.

CALAND, W. 1901. *De Wenschoffers.* Amsterdam.

CAMERON, A. L. P. 1885. 'Notes on some tribes of New South Wales'. *JAI*, 14, pp. 344–70.

CHEVILLARD, SIMILIEN. 1889. *Le Siam et les Siamois.* Paris.

Contributions to North American Ethnology 1877–. Washington.

CURR, EDWARD M. 1886–7. *The Australian Race.* (4 vols.) Melbourne.

CUSHING, FRANK HAMILTON. 1883. 'Zuñi fetishes'. *ARBE*, 2, pp. 9–45.

   1896. 'Outlines of Zuñi creation myths'. *ARBE*, 13, pp. 321–447.

DEUSSEN, PAUL. 1894–1917. *Allgemeine Geschichte der Philosophie.* Leipzig.

DOOLITTLE, JUSTUS. 1876. *Social Life of the Chinese.* New York.

89

Bibliography

DORSEY, J. OWEN. 1884. 'Omaha sociology'. *ARBE*, 3, pp. 205–370.

1894. 'A study of Siouan cults'. *ARBE*, 11, pp. 351–544.

1896. 'Siouan sociology'. *ARBE*, 15, pp. 205–244.

DURKHEIM, ÉMILE. 1898. 'La prohibition de l'inceste'. *Année Sociologique*, 1, pp. 1–70.

EBERS, GEORG. 1901. 'Die Körpertheile, ihre Bedeutung und Namen in Altägyptischen'. *Abhandlungen der Königlich Bayerische Akademie der Wissenschaften*, 21, pp. 79–175.

FEWKES, JESSE WALTER. 1897. 'Tusayan fetishes'. *ARBE*, 15, pp. 245–313.

FISON, LORIMER and HOWITT, A. W. 1880. *Kamilaroi and Kurnai*. London.

FLETCHER, ALICE C. 1898. 'The significance of the scalp-lock: a study in Omaha ritual'. *JAI*, 27, pp. 436–50.

FRAZER, J. G. 1887. *Totemism*. Edinburgh.

1899. 'The origin of totemism'. *Fortnightly Review*, 65, pp. 647–65, 835–52. London.

GROOT, JOHANN JACOB MARIA DE. 1892–1910. *The Religious System of China*. (6 vols.) Leiden.

HADDON, A. C. 1890. 'The ethnography of the western tribe of Torres Straits'. *JAI*, 19, pp. 297–440.

1901. *Head-hunters, Black, White and Brown*. London.

HILLEBRANDT, ALFRED. 1897. *Ritual-Litteratur: Vedische Opfer und Zauber*. Leipzig.

HOWITT, A. W. 1883. 'Notes on the Australian class systems'. *JAI*, 12, pp. 496–510.

1884a. 'On some Australian beliefs'. *JAI*, 13, pp. 185–98.

1884b. 'Remarks on the class systems collected by Mr. Palmer'. *JAI*, 13, pp. 335–46.

1884c. 'On some Australian ceremonies of initiation'. *JAI*, 13, pp. 432–59.

1885. 'Australian group relations'. *Report of the Regents of the Smithsonian Institution, 1883*, pp. 797–824. Washington.

1886. 'On the migration of the Kurnai ancestors'. *JAI*, 15, pp. 409–21.

1887. 'On Australian medicine men: doctors and wizards of some Australian tribes'. *JAI*, 16, pp. 23–58.

1889a. 'Further notes on the Australian class systems'. *JAI*, 18, pp. 31–68.

1889b. 'Notes on Australian message sticks and messengers'. *JAI*, 18, pp. 314–32.

HUNT, ARCHIBALD E. 1899. 'Ethnographical notes on the Murray Island, Torres Straits'. *JAI*, 28, pp. 5–19.

KOHLER, J. 1897. *Zur Urgeschichte der Ehe.* Stuttgart.

LA GRASSERIE, RAOUL DE. 1899. *Des Religions comparées au point de vue sociologique.* Paris.

LA LOUBÈRE, SIMON DE. 1714. *Description du Royaume de Siam.* Amsterdam.

LANG, ANDREW (trans. MARILLIER, L.). 1896. *Mythes, Cultes, et Religions.* Paris.

LECLÈRE, ADHÉMAR (ed.). 1898. *Les Codes cambodgiens.* Paris.

LEGGE, JAMES (trans.). 1882. *The Yî King.* (Sacred Books of the East, vol. 16.) Oxford.

LUMHOLTZ, CARL. 1900. *Symbolism of the Huichol Indians.* New York.

MACCAULEY, CLAY. 1887. 'The Seminole Indians of Florida'. *ARBE*, 5, pp. 469–531.

MACGEE, W J 1894. 'The Siouan Indians'. *ARBE*, 15, pp. 157–204.

MALLERY, GARRICK. 1886. 'Pictographs of the North American Indians'. *ARBE*, 4, pp. 3–256.

MATHEWS, R. H. 1895. 'The Bora or initiation ceremonies of the Kamilaroi tribe'. *JAI*, 24, pp. 411–27.

1896. 'The Bora or initiation ceremonies of the Kamilaroi tribe' (cont.). *JAI*, 25, pp. 318–39.

1898. 'Divisions of Australian tribes'. *Proceedings of the American Philosophical Society*, 37, pp. 151–4. Philadelphia.

1900. 'The Wombya organisation of the Australian aborigines'. *American Anthropologist*, 2, pp. 494–501.

MATTHEWS, WASHINGTON. 1887. 'The Mountain Chant: a Navajo ceremony'. *ARBE*, 5, pp. 379–467.

MINDELEFF, VICTOR. 1891. 'A study of Pueblo architecture, Tusayan and Cibola'. *ARBE*, 8, pp. 3–228.

Bibliography

MORGAN, LEWIS H. 1851. *League of the Ho-dé-no-sau-nee, or Iroquois*. Rochester, N.Y.
1877. *Ancient Society*. New York.

MOURA, JEAN. 1878. *Vocabulaire français-cambodgien et cambodgien-français*. Paris.

MÜNSTERBERG, HUGO. 1889–92. *Beiträge zur experimentellen Psychologie*. Freiburg-im-Breisgau.

NEGELEIN, JULIUS VON. 1901. 'Die Volksthümliche Bedeutung der weissen Farbe'. *Zeitschrift für Ethnologie*, 33, pp. 53–85.

PALLEGOIX, JEAN-BAPTISTE. 1854. *Description du Royaume thaï ou Siam*. (2 vols.) Paris.
1896. *Dictionnaire siamois-français-anglais*. Bangkok.

PALMER, E. 1884. 'Notes on some Australian tribes'. *JAI*, 13, pp. 276–334.

PETITOT, ÉMILE. 1887. *Traditions indiennes du Canada nord-ouest*. Alençon.

PFEIL, JOACHIM VON. 1899. *Studien und Beobachtungen aus der Südsee*. Braunschweig.

POWELL, J. W. 1881. 'Wyandot government: a short study of tribal society'. *ARBE*, 1, pp. 57–69.
1883. 'Report of the Director'. *ARBE*, 2, pp. xv-xxxvii.
1896. 'Administrative report'. *ARBE*, 13, pp. xxi-lix.

RIGGS, STEPHEN RETURN. 1883. *Tah-koo Wah-kan: the Gospel among the Dakotas*. Washington.

RIVERS, W. H. R. 1900. 'A genealogical method of collecting social and vital statistics'. *JAI*, 30 (n.s. 3), pp. 74–82.

ROTH, W. E. 1897. *Ethnological studies among the northwest-central Queensland Aborigines*. London.

SMITH, E. A. 1883. 'Myths of the Iroquois'. *ARBE*, 2, pp. 47–116.

SMYTH, ROBERT BROUGH. 1878. *The Aborigines of Victoria*. London.

SPENCER, B. and GILLEN, F. J. 1899. *Native Tribes of Central Australia*. London.

STEINEN, KARL VON DEN. 1894. *Unter den Naturvölkern Zentral-Brasiliens*. Berlin.

Bibliography

STEVENSON, MATILDA COXE. 1894. 'The Sia'. *ARBE*, 11, pp. 3–157.

USENER, HERMANN. 1887. *Epicurea*. Bonn.
 1896. *Götternamen: Versuch einer Lehre von der Religiösen Begriffsbildung*. Bonn.
 1898. 'Göttliche Synonyme'. *Rheinisches Museum für Philologie*, 53, pp. 329–379.

WILLIAMS, SAMUEL WELLS. 1899. *The Middle Kingdom*. New York.

YOUNG, ERNEST. 1898. *The Kingdom of the Yellow Robe*. London.

INDEX